Heredity under the Microscope

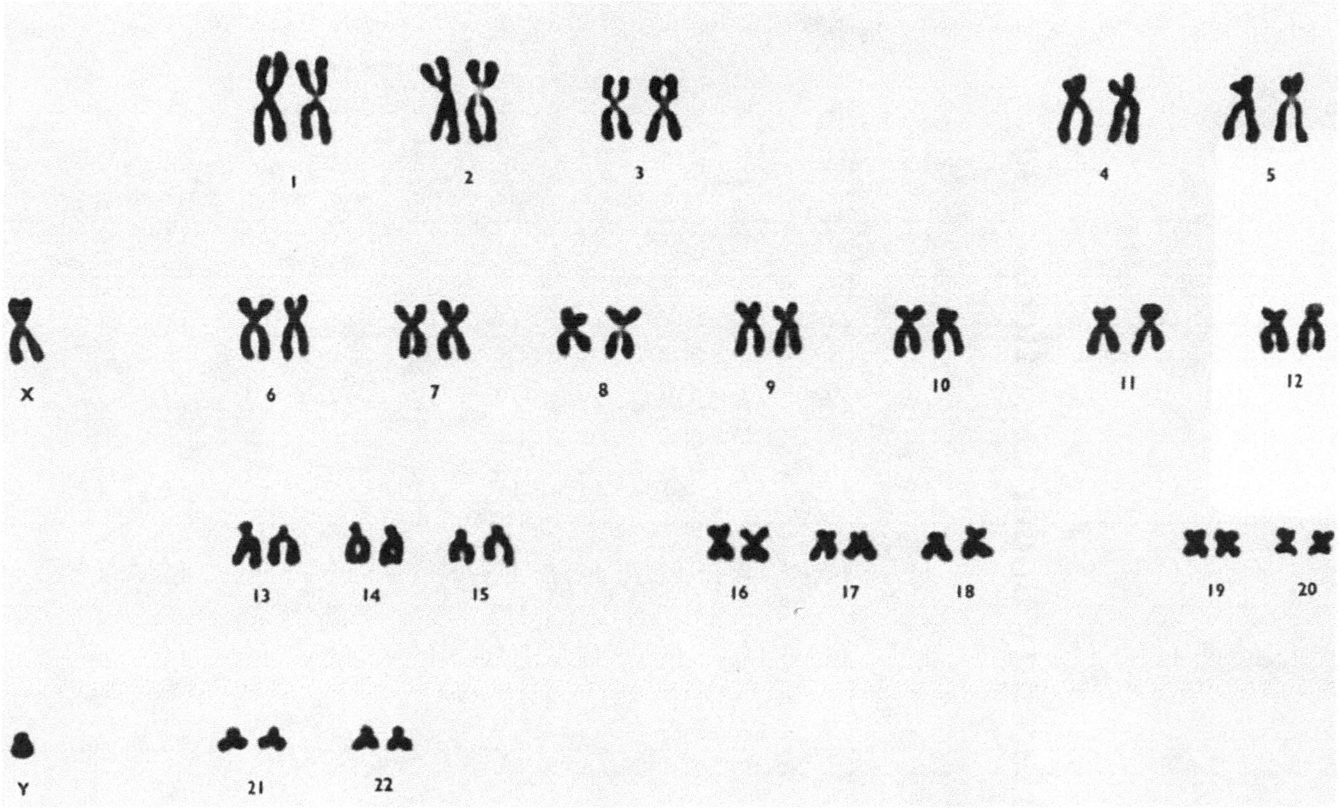
1
2
3
4
5
X
6
7
8
9
10
11
12
13
14
15
16
17
18
19
20
Y
21
22

Heredity under the Microscope

Chromosomes and the Study of the Human Genome

SORAYA DE CHADAREVIAN

THE UNIVERSITY OF CHICAGO PRESS   CHICAGO AND LONDON

The University of Chicago Press, Chicago 60637
The University of Chicago Press, Ltd., London
© 2020 by The University of Chicago
Published 2020

29 28 27 26 25 24 23 22 21 20   1 2 3 4 5

ISBN-13: 978-0-226-68508-3 (cloth)
ISBN-13: 978-0-226-68511-3 (paper)
ISBN-13: 978-0-226-68525-0 (e-book)
DOI: https://doi.org/10.7208/chicago/9780226685250.001.0001

Frontispiece: Human karyotype (1960). (Human Chromosomes Study Group, "Proposed Standard of Nomenclature," 5. Reproduced with permission of Mac Keith Press, London.)

Library of Congress Cataloging-in-Publication Data

Names: Chadarevian, Soraya de, author.
Title: Heredity under the microscope : chromosomes and the study of the human genome / Soraya de Chadarevian.
Description: Chicago ; London : The University of Chicago Press, 2020. | Includes bibliographical references and index.
Identifiers: LCCN 2019052054 | ISBN 9780226685083 (cloth) | ISBN 9780226685113 (paperback) | ISBN 9780226685250 (ebook)
Subjects: LCSH: Chromosomes—Research—History—20th century. | Chromosomes.
Classification: LCC QH600 .C45 2020 | DDC 572.8/7—dc23
LC record available at https://lccn.loc.gov/2019052054

The chromosomes, we declare, are the little things that make us what we are.

DARLINGTON, "The Chromosomes as We See Them"

I suppose that my real favorites are the chromosomes, whose ever-enchanting beauty is addictive to some microscopists, including myself.

HSU, "My Favorite Cytological Subject"

Contents

Introduction

In the 1960s and well beyond, pictures of orderly paired chromosomes were the most iconic images of genetics. They appeared in clinical records, on the pages of newspapers, in courtrooms, and on greeting cards, with the chromosomes serving as genetic "portraits" and providing insights into the inner "self" of individuals.[1] By the end of the decade, the prominent British human geneticist Lionel Penrose announced that the study of human chromosomes, which, until recently, had been almost "completely unexplored territory," had become "a happy hunting ground for thousands of investigators all over the world."[2] In the vision of its promoters, the techniques for analyzing chromosomes had wide implications for the study of a growing number of genetic diseases and mental conditions; for the study of cancer, the biology of sex determination, infertility, and aging; for epidemiological investigations and comparative studies of human populations; in radiation studies and toxicology; in the courts; and in the policy arena.

Surprisingly, the microscopic study of human chromosomes took off at exactly the time when molecular approaches to heredity were celebrating their biggest advances. The suggestion that humans usually have forty-six rather than forty-eight chromosomes as had been the orthodoxy for many years, followed on the heels of the proposal of the double helical structure of DNA. Scientists celebrated the consensus on the new number of human chromosomes as the beginning of a "new era" in the study of human heredity.[3] Yet historical accounts often draw a direct line from the double helix to the genome sequencing projects of the 1990s, without much reference to the chromosome studies of the intervening decades. What was behind the explosive growth in human chromosome research? And how can we explain its paradoxical place in the history of human heredity? *Heredity under the Microscope* sets out to answer these questions. Taking as its focal point chromosomes and the techniques and images that come packaged with them, it traces the expanding uses of these genetic tools and the questions and concerns that propelled them. It aims to provide an integrated account that makes space for microscope-based practices next to molecular approaches in the quest to study and harness heredity in humans.

Much of the fascination with chromosomes and the persuasive power of the work was based on the visual evidence the chromosome preparations provided. Critics contended that looking at pictures was not enough to understand the mechanisms at work. In focusing on the visual practices that sustained work with chromosomes, this book argues that the patient collection of cases and the often bewildering variety of observations made by chromosome researchers looking down the microscope were as central to the making of human genetics as was the search for molecular mechanisms gleaned from the study of simple organisms pursued at the same time.

Chromosomes and the Study of Human Heredity

The study of human chromosomes was not new in the postwar era.[4] Observation under the microscope of the strongly stained bodies (or "chromo-somes," from the Greek for *color* and *bodies*) in the cell nucleus and their identification as the hereditary material goes back to the late nineteenth and early twentieth centuries. The chromosome theory of heredity remained contested, but chromosomes meanwhile became the object of extensive research. Armed with much improved compact multilens microscopes, botanists and zoologists studied the ordered movements, or "dance of chromosomes," during ordinary cell division (mitosis) and in the formation of reproductive cells (meiosis).[5] They established that each plant and animal species had a fixed and characteristic number of chromosomes in every cell. They compiled lists and produced atlases comparing the number of chromosomes in various species throughout the plant and animal kingdoms.[6] In the 1920s scientists agreed that humans (including "whites" and "Negroes") had forty-eight chromosomes, an observation often confirmed over the years. The same number was counted in Rhesus monkeys, whereas Capuchin monkeys had fifty-six chromosomes, giving rise to speculations about the evolutionary mechanism behind the numbers.[7] Yet human chromosomes, like chromosomes from mammalian cells more generally, were difficult to work with, not least because there are so many of them. Fruit flies have only four chromosomes and onions have eight. In addition, access to human tissue suitable for chromosome analysis was anything but straightforward. To find dividing cells in which chromosomes could be studied, tissue from testes was the preferred material. As one protagonist laconically remarked, for "the knights of the dark ages" of cytogenetics, there were only two ways to obtain such samples: "waiting outside the operating rooms and waiting by the gallows."[8] For these reasons—and for the potential practical use in crop breeding—most research on chromosomes was performed on insect or plant cells.[9]

The extensive effort in Thomas Hunt Morgan's laboratory at Columbia in the 1910s and 1920s to map genes in the fruit fly according to their relative position on the chromosomes eventually confirmed the chromosome theory of heredity. The giant chromosomes in the salivary glands of the fly larvae made genetic activity visible under the microscope. Together with the chromosome map that was being generated and the extensive collection of mutant flies that was built up parallel to it, this tool established the fruit fly as the organism of choice for genetic research.[10] For decades the fly remained a point of reference for much genetic work done on other organisms, including humans. Yet after World War II, widespread efforts to establish the effects of nuclear radiation in humans as well as a continuing interest in the role of chromosomes in the etiology of cancer—a disease increasingly linked to the risks of radiation—provided new incentives to develop better protocols for studying human chromosomes. The close connections between genetics and the atomic age have been explored before.[11] However, Cold War anxieties about the effects of atomic radiation played a particularly salient role in human chromosome research, as the new preparation techniques promised to make mutations directly visible under the microscope. It was in this context of increased concerns around mutations and human heredity that the recount of human chromosomes took place.[12]

After the new chromosome count was settled in the late 1950s, research on human chromosomes entered the period of explosive growth described by Penrose. In the course of a few years, the study of human chromosomes became the most dynamic area of chromosome research.[13]

Skills and observations gained in the study of chromosomes in plants, the fruit fly, and the mouse remained important and traveled to and fro among different research communities. Nevertheless, the focus of this book is on human chromosomes because heredity in humans raised a specific set of questions, and its role in the explanation of human behavior remained deeply contested in the middle decades of the twentieth century. At the same time, research on human chromosomes contributed decisively to pushing the study of human heredity to the forefront of research. The aim is not a systematic history of human cytogenetics, the field concerned with the study of human chromosomes.[14] In fact, taking human chromosomes and the techniques and images that travel with them as the focal point of study makes it possible to transcend the boundaries of a disciplinary history and trace the many contexts in which the techniques were taken up. In this study human chromosomes serve as an analytical lens to gain insight into where human heredity mattered and genetic knowledge was embraced, debated, or rejected.[15]

Human chromosome research was just one of many approaches to the study of human heredity. In the same text mentioned earlier, Penrose distinguished cytogenetics (based on microscopic observation) from "classical human genetics" (based on the construction of pedigrees) and "established quantitative and biochemical methodologies." Cytogenetics, he concluded, raised new questions that could not be solved by "preconceived or routine ideas."[16] A World Health Organization technical guide listed a battery of methods for the genetic study of human populations, including the determination of blood groups and immunological and biochemical markers such as leukocyte antigens and phenylthiocarbamide (PTC) tasting, whose distributions vary in different populations.[17] In particular, the study of blood groups had long played a key role in the study of human heredity.[18] Yet proteins and antigens, including the complex ABO system, provided only indirect proof of variation on the genetic level. Moreover, they offered insights into the genetic variation of just one factor. In contrast, chromosome preparations—at a glance—offered a picture of the whole genome, a term introduced in the 1920s to denote the complete (single) chromosome set of an individual or a species. In addition, the same techniques could be used to study hereditary or congenital mutations as well as mutations accumulated in the lifetime of individuals, in somatic as well as reproductive cells. For these reasons, chromosome analysis promised to be a much more powerful tool for the study of human genetics. Following chromosomes, then, makes it possible to recover a broad range of preoccupations around human heredity and to track the expanding uses of genetic techniques, the visual evidence they provided, and their meanings. Today we tend to think of chromosome analysis as predominantly a diagnostic tool. Yet this was not always so, and we need both to ask how these practices became entrenched in the clinic and to tend to the other uses that shaped the technology and discussions around human heredity more broadly.

Recovering the multiple contexts in which chromosomes were embedded also helps disentangle the study of postwar human heredity from the predominant concern about continuities with eugenic practices. The exclusive concentration on the "eugenic question," as important as it is, obscures other aspects of the postwar study of human heredity and its many ramifications in science, medicine, and politics.[19] The loaded questions of "who should and who should not inhabit the world" and who decides continued to vex proponents and critics of chromosome techniques. They gained special significance in the context of prenatal diagnosis that became more widely available in the 1970s.[20] Yet prenatal diagnosis was only one of a wide range of issues tackled with the new techniques. Chromosome researchers for their part rejoiced that the new cytogenetic techniques put the study of human heredity on a "very solid basis," distancing it from the speculative approaches of the past.[21]

Making Visible and Seeing

The questions remain: What are chromosomes? And what does it mean to treat them as visual objects?

Every cell of the human body (with the exception of red blood cells) has a full set of chromosomes.[22] Yet chromosomes become visible only through sustained intervention and skilled observation.[23] The techniques employed to render chromosomes visible under the microscope have been molded and transformed through time. The exact protocols differ locally and depend on the specific aims of the analysis, but the preparation always demands complex, precisely timed routines, concentrated human attention, and skill, even with the advent of increasing automation. The work involved in producing the pair of "classic" photographs that accompanied the article in which Joe Hin Tjio and Albert

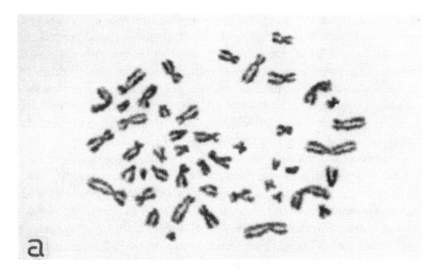

FIGURE I. Photomicrograph of human chromosomes published by Tjio and Levan in 1956. The authors counted forty-six chromosomes. The photograph showed "the ease with which the counting could be made" (p. 2).

Source: Tjio and Levan, "Chromosome Number of Man," 2, fig. 1a. Reproduced with permission of *Hereditas*.

Levan, of the Cancer Chromosome Laboratory at the University of Lund, first suggested that humans have forty-six rather than forty-eight chromosomes provides a useful example (fig. 1).

To prepare the chromosomes for observation, the two researchers tinkered with a set of newly available techniques. Departing from the practice of using embedded tissue blocks from which thin sections were cut, they started from cell cultures of embryonic tissue that grew in a thin layer. The cultures were provided to them by a colleague in the Virus Laboratory in Lund who had access to human embryos from legal abortions. Once the cultures had grown, Tjio and Levan treated the cells with colchicine, a cell poison that interferes with the formation of the cell spindle that pulls sister chromatids apart during cell division to distribute them into the two daughter cells. As a consequence, cell division is interrupted at the stage known as metaphase, when the diffused, double-up chromatin fibers are condensed into compact chromosome structures and become visible under the microscope. The characteristic X-shaped form seen on Tjio and Levan's photos and other chromosome images from the time is an artifact of colchicine treatment. Subsequently, the two researchers added a hypotonic solution consisting of a balanced salt solution mixed with distilled water to the culture medium to swell the cells and separate the chromosomes. They stained the cells with a dye that specifically binds to chromosomes, placed the preparations on a glass slide, and carefully pressed the covers with their thumbs in a final attempt to spread the chromosomes into a two-dimensional plane. The chromosome preparations could then be viewed under a conventional light microscope and drawn with the help of a camera lucida as well as photographed.

The epistemic value of drawing and photography was a matter of dispute between the two authors of the paper. For Levan, seeing was intimately connected to the act of drawing. In contrast, Tjio dismissed drawing as a subjective way of interpreting what one sees under the microscope and instead invoked the power of photography to record the microscopic image and provide the decisive evidence. Yet far from relying on the "mechanical objectivity" of photography, he further manipulated the images in the dark room by applying all the tools available to produce the high-quality prints that so impressed fellow researchers, including Levan, his reservations against photography notwithstanding.[24] The dispute between Levan and Tjio underlines the importance and contested nature of visual evidence in work on chromosomes. With the preparations improving and the chromosomes showing fewer overlaps that needed to be resolved, photography gained the upper hand over drawing in chromosome laboratories. Yet resistance against photographic techniques persisted, and some laboratories preferred to count chromosomes under the microscope rather than from photographs. Against the suggestive power of photomicrographs, the philosopher of science Ian Hacking reminds us that "the reality in which we believe is only a photograph of what came out of the microscope, not any credible real, tiny thing."[25] Nevertheless, the visual evidence of chromosome images relied on the sedimented experience of working with microscopes and the researchers' familiarity with analyzing the subcellular world to which microscope preparations provided access.[26]

Hans-Jörg Rheinberger, reflecting on the intersection between instruments and biological objects—or, in his terms, on "the reciprocal relation between epistemic things and the technical conditions of their manipulation in experimental systems"—has remarked on how objects must be configured "in such a way that these instruments can do their job."[27] Specifically with respect to microscopic preparations, the microscope demands that "objects be presented in a flat, two-dimensional form, since microscopes produce a sharp image only in the focal plane."[28] A further characteristic of microscopic preparations is that "as a rule things tailored to the lens cannot themselves be seen during the process of preparation. . . . The process of their production escapes the eye: only the gaze through the microscope decides after the fact whether it was successfully carried out. This requires that the preparer focus on the regularities of the production process, which must so to speak function blindly."[29] This characterization fits perfectly with the routines—from the use of tissue cultures that grow in one layer and the final squashing of the cells on the slide to the rendering of the microscope image on a two-dimensional plane (paper, photograph, or screen) and the minutely followed protocols for the preparation of high-quality chromosome spreads—developed around the visualization of chromosomes. Summing up, we can say that chromosomes that researchers see through the eyepiece of a microscope, on a photograph captured from the visual field of a microscope, or more recently on the computer screen are microscopic objects, not just in the sense that they can be seen only through a microscope but also in the sense that they are prepared in such a way that they *can* be viewed under the microscope.[30]

Photographs (like drawings in other ways) captured the microscope image and preserved the experimental evidence. They provided proof and invited other scientists to check the evidence. Discussions of chromosome counts took place around photographic images. Nevertheless, chromosomal observation required extensive training, and even such a seemingly basic activity as counting chromosomes was anything but straightforward.[31] Photographs themselves became the object of further manipulation. Enlarged prints were cut up and the chromosomes ordered according to a standardized scheme agreed on in specially convened standardization conferences.[32] Photographic slides could be projected against a screen, facilitating the measurement of the magnified chromosomes. Thus, through photography chromosomes became tangible objects that could be further manipulated, measured, and sorted, even if in the service of identifying and counting chromosomes other information on the place and function of chromosomes in the cell was lost. Indeed, during preparation researchers took great care to keep the complete set of chromosomes of each cell together, but the contours of the nuclear structure that contained the chromosomes disappeared from the photographic images. This aided the perception that chromosomes could be studied as independent objects, isolated from their milieu (fig. 2).

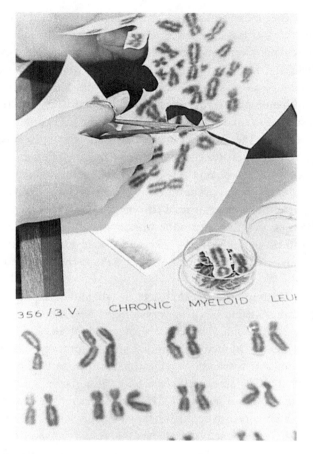

FIGURE 2. Cutting out chromosomes.
Source: WHO Archives, Album EURO-RESEARCH, WHO/9258, photo library reference WHO_A_021972. © World Health Organization/Spooner, 1963. Reproduced with permission.

Over the years the protocols for preparing human chromosomes for microscopic observation were constantly refined and updated. From the 1960s, most work was performed on white blood cells, isolated from small samples of peripheral blood and grown in culture. The new technique greatly facilitated access to human material for chromosome analysis. Squashing was replaced by air-drying, which produced a similar flattening effect while requiring less skill to apply. Chromosomes were banded with fluorescent and other stains or variously labeled, first with radioactive and later with genetically engineered fluorescent markers. The techniques revealed a host of new details and quickly made previous approaches look "antiquated."[33] Colchicine treatment was adapted to produce short or less condensed longer chromosomes, showing more bands. The changes went hand in hand with the introduction of ever more refined light and later fluorescence microscopes with in-built cameras and digital image processing. Analysis moved from counting to close observation of banding patterns, using visual standards. By the 1970s, automation started to take over some of the routine tasks of chromosome analysis, but even the most recent chromosome-sorting software programs still rely on extensive checking by highly skilled human observers. Today, much work is done on the computer screen, but whenever a doubt arises, the slide with the preparation is placed under a microscope for direct inspection. Training to read chromosome images takes one year. More time is needed to perform the work confidently. Not everyone has the ability to become a skilled observer. Women are generally considered more adept at this than men, which speaks to the gendered division of labor in the laboratory.[34]

Lorraine Daston has suggested that visualization techniques do not just make things visible but also "crystallize" new objects of scientific inquiry. She speaks of an "ontology wrought by observation."[35] Seeing here always means "trained perception" and is a practice that is shared by a community of researchers.[36] Daston also points to the "aesthetic pleasures of skillful perception," of seeing "at a glance" or the "all-at-once-ness" of skillful observation.[37] Her examples mostly stem from the eighteenth and nineteenth centuries, but she could just as well be speaking of the visual experiences of chromosome researchers in the 1950s. In his personal account of human chromosome research, Tao-Chiuh Hsu, a pioneer in the field, described chromosomes as "hypnotically beautiful objects," even comparing them with Rembrandt paintings.[38] Forty years later, a still-practicing cytogeneticist echoed Hsu's views when she confessed, "Were it not for the esthetic appeal of chromosomes, I would have left the field some time ago." Elaborating on her statement, she added: "I wonder what Levan would have thought of the current 'imaging systems,' which allow for much more manipulation of 'captured images' (the term used) than photography. I am one of a dying breed of cytogeneticists who actually still prefer to look through the microscope because so much of the texture and depth of the chromosomes is lost in translation to 2D images. Granted, the chromosomes have been manipulated (spread, fixed, stained etc.) before viewing at high magnification but a look through a microscope ocular is very different from a look at a computer screen or sheet of paper."[39] Chromosomes for cytogeneticists thus were and still are tightly bound to and indeed inseparable from artisanal and visual practices, including cell preparation and staining techniques, trained microscopic observation, drawing, photographing, and digital displays. Counting, measuring, and mapping chromosomes happened around visual representations. Chromosomes also became mobile and traveled with these images and the practices on which they depended. Drawings, photographs, or diagrammatic arrangements of chromosomes regularly accompanied scientific publications. In clinical atlases, chromosome pictures with arrows pointing to anomalies in the number or form of certain chromosomes were paired with photographs of anonymized and objectified patients showing specific morphological or behavioral characteristics. Projected in lecture halls, the chromosome images impressed scientific audiences. At the same time, chromosome pictures also became recognizable images for a wider public, appearing in clinical settings, in media reports, and even providing the pattern for a Marimekko fabric print. Following chromosomes, then, means attending to the wide gamut of visual practices that sustained them and around which contentions took shape. The reliance on visual evidence represented the strength but also the weakness of chromosome research, especially in the eyes of molecular biologists who spurned images in favor of mathematical formulations and causal explanations.

Visualization became the defining criterion to demarcate what belonged within the field and what fell outside it. Molecular labeling techniques such as fluorescent in situ hybridization (FISH) were readily integrated into the tool kit of cytogeneticists as long as they served to make chromosomal structures visible under the microscope.[40] With current molecular practices increasingly relying on microscopic imaging, this demarcation is becoming less sharp, an issue that will become significant for the argument of the book.

New techniques—from the use of blood cultures to the various banding techniques and automation that allowed researchers to simplify and speed up or bring more intricate structural details into view—kept interest in chromosome analysis alive. At the same time, chromosome techniques and images were enrolled in an ever-expanding series of projects that, in turn, changed what chromosomes were about and where human heredity mattered. Mapping their various uses, we see chromosomes becoming objects of research, entering the clinic and turning up in patient records, becoming instruments for surveillance and tools to measure exposure to radiation and other environmental toxins, being employed to test gender and define identities, appearing in court records and in deliberations on policy and law, becoming a matter of dispute and ethical debate. Through these appropriations and contentions the meanings of chromosomes and of human heredity expanded and changed.

Following Chromosomes

Heredity under the Microscope traces the history of human chromosome research from its rise to prominence in the 1950s to the 1980s, when the mapping of the human genome was in full swing and molecular technologies started to compete directly with cytogenetic approaches. It reconstructs the political and scientific concerns that propelled the study of human chromosomes and investigates the many fields—from radiobiology and cancer research to medical genetics, gender testing, criminology, and the genetic study of human populations—where techniques for studying chromosomes made an entry, providing new answers to existing questions and opening up new areas of investigation and debate.

Historians have written about the strength of the British school of human genetics from the 1930s into the postwar years, with the Galton Laboratory, headed first by Ronald A. Fisher and then Penrose, forming an important hub.[41] Partly building on this tradition, in the 1950s, Britain also became a hotbed for research on human chromosomes. Playing pivotal roles were the Radiobiological Research Unit at the British Atomic Energy Research Establishment in Harwell, one of the two key sites of the British atomic bomb project, and the Medical Research Council Clinical and Population Cytogenetics Unit, headed by Michael Court Brown in Edinburgh, next to other centers in London, Oxford, and Glasgow. This is also where research for this study started. Work performed at the Edinburgh unit, especially, provided material for various topics discussed in this book. However, even if the story often circles back to some of the early work and debates on chromosomes in Britain, the story told here is not British. The study draws attention to the small international group of researchers stemming from Sweden, the United Kingdom, France, Japan, and the United States that together pioneered the use of human chromosome analysis in the mid-1950s and early 1960s. It highlights the role of international organizations such as the World Health Organization in promoting the use of chromosome techniques for the study of effects of global fallout, for clinical studies, and for worldwide population studies. The effort to provide a chromosome map of all human genes, a goal that chromosome researchers pursued intensely from the 1970s, also built on international cooperation. More generally, chromosome techniques were versatile and mobile, but not effortlessly so. Following chromosomes thus requires attending to the ways that chromosome techniques and samples traveled and were picked up in laboratories and clinics around the world. Mapping the expanding meanings of chromosomes while teasing out their epistemic role as visual objects, the book argues that the study of human chromosomes and microscopic work were as central to postwar concerns around heredity as the biochemical and molecular approaches that flourished at the same time.

Chapter 1 substantiates the claim that anxieties of the atomic age were the driving force for a new interest in human chromosomes. By providing a technique to visualize mutations, chromosome analysis emerged as the "right tool" at the right time to address a host of urgent political and scientific questions raised by the development of atomic radiation for civilian and military uses.[42] Concerns surrounding radiation and other pollutants remained at the center of much chromosome research throughout the 1950s and 1960s and provided new legitimization for human heredity research that had been discredited by its implication in eugenic and racial practices. Yet human chromosome research and, with it, questions around human heredity also expanded into new areas. These are explored in chapters 2 to 4, which highlight three interlocking thematic fields—the clinical career of chromosomes, the study of sex and crime, and the genetic study of human populations. Together the chapters demarcate the large territory in which human chromosomes and with them genetic explanations came to matter.

Having mapped the scope of human chromosome studies in the postwar era, chapter 5 addresses the relations between microscope-based and molecular approaches to heredity. In particular, it reconsiders what is often described as the "molecularization" of chromosome research in the light of the simultaneous turn of molecular biology to human and medical genetics, a field long occupied by chromosome researchers. This chapter also expands the analysis more decidedly into the 1970s and beyond. The epilogue reflects on the current resurgence of interest in the architecture, spatial distribution, and regulatory functions of chromosomes and further examines the role of visual evidence in staking knowledge claims. Considering current directions in the life sciences, it makes the case that chromosome research was not just "old-fashioned" biology that was superseded by molecular approaches but that it made its own distinct contributions to the study of the human genome and its continued, if contested, salience, then and today.

CHAPTER ONE

Radiation and Mutation

The history of human chromosome research has often been told as a history of sometimes fortuitous, other times hard-won technical improvements in the art of preparing and analyzing chromosomes. In the early 1950s, cell-culturing techniques and the pretreatment of tissue with hypotonic medium that swelled the cells and spread apart the chromosomes did away with the need to work from serial sections in which the chromosomes were clumped together and sliced up. The improved preparation techniques inspired new work with human chromosomes. Together with the use of colchicine and the development of squash techniques, the number of human chromosomes was revised and modern cytogenetics began. The introduction a decade later of staining techniques that made it possible to distinguish every single chromosome by its characteristic banding pattern made everything that had come before appear "paleolithic," and cytogenetics came into its own until molecular technologies and new fluorescent marking techniques once more dramatically increased the resolution of chromosomal observation.[1]

Yet this story, as close as it brings us to the laboratory bench, begs the following questions: What attracted researchers to the study of human chromosomes, and what sustained their interest? What gave importance to the observations they made?

Postwar genetics was deeply intertwined with the challenges and opportunities of the atomic age. This holds true specifically for human chromosome research. If we search for the atomic connections, they are pervasive and deeply mark the history of the field. Efforts to establish the effects of radiation in humans, along with renewed interest in the role of chromosome mutations in causing cancer—a disease often linked to radiation exposure—provided new incentives to develop methods to study human chromosomes at a time when various countries were developing atomic energy for military and civilian uses. Many of the decisive preparation techniques for human chromosome analysis (or karyotyping) were originally devised to study the chromosomes of humans or experimental animals with radiation-induced leukemia, a cancer of the white blood cells also dubbed the "pestilence of the atomic age."[2] This was true for both the bone marrow method developed by Charles Ford and Patricia Jacobs at the Radiobiological Research Unit at Harwell, Britain's Atomic Energy Research Establishment, in the mid-1950s, as well as for the peripheral blood method developed by Peter Nowell and David Hungerford in Philadelphia that soon replaced it.

Similarly, the researchers involved in developing and promoting human chromosome analysis in the middle decades of the twentieth century were deeply involved in things nuclear. They worked in institutions or projects funded to assess the effects of radiation. They visited atomic bomb explosion sites in Japan to study the survivors and test sites in the Pacific to record and plan experiments. They sat on numerous national and international committees dealing with the effects of radiation on humans, wrote reports to their respective governments, suggested "permissible doses" of radiation based on what they knew was incomplete knowledge, and forcefully argued for the urgent need to expand genetic research to increase knowledge on the genetic structure of human populations and so help assess the effects of radiation.[3] The topics they tackled included the effects of radiation treatment in the clinic, studies of atomic bomb survivors, and the long-term effects of low-dose radiation exposure on the workplace or from fallout. Funds were forthcoming, and policy makers, the media, and the public eagerly received the results.

This chapter substantiates the claim that concerns of the atomic age provided tools, urgency, and visibility to human chromosome research. It traces the intimate connection of chromosome research with efforts to capture the effects of atomic radiation in humans in the aftermath of the atomic bombings in Japan and in the face of the continuing development of atomic energy for military and civilian uses. It follows the careers of key postwar protagonists of human chromosome research and reflects on the sites and resources of their work. The chapter then takes a closer look at two lines of research that defined cytogenetic research agendas while directly responding to atomic age concerns: the chromosomal study of various forms of leukemia and the use of chromosomes as a tool to measure exposure to radiation and other workplace and environmental pollutants. At the basis of both these endeavors was the aim to visualize mutations and to study and monitor their effects. The term *atomic age* is here used to denote the first decades following World War II—a time when nuclear politics was dominating many aspects of public life, from military strategies to foreign policy and the economy, from national research agendas to public debates.[4]

Visualizing Mutations

The mutational effect of radiation literally exploded as an issue with the dropping of the two atomic bombs on the populated cities of Hiroshima and Nagasaki that marked the end of World War II. As historians have argued, the atomic bomb was conceived as a super-explosive rather than an atomic weapon in the more literal sense.[5] People exposed to the effects of the bomb were expected to die, not to survive the explosion and continue to suffer from the lingering effects of radiation.[6] The first images that reached the public showed the material devastation produced by the bomb but did not reflect the plight of the survivors. The harrowing pictures of the "walking dead" taken by photojournalist Yoshito Matsushige on the day the bomb fell on Hiroshima, for instance, were published only after the end of the American occupation.[7] Nonetheless, the radiation effects of the two bombs on the surviving population quickly became a medical, diplomatic, political, and ethical problem of vast dimensions that was further complicated by the postwar development and testing of atomic weapons and rising Cold War tensions.[8]

Susan Lindee has described how the genetic effects of the bomb moved to center stage in the work of the Atomic Bomb Casualty Commission. Apparently, this was largely because of the interest of the leading scientist on the American team, James Neel. The commission based its first assessment of the genetic effects of the atomic bombings on such indicators as the rates of stillbirths, sex ratio, congenital anomalies, infant mortality, bodily dimensions, and life span in the children of the survivors. All these factors were considered related to the mutational effects of radiation. Lindee has discussed the problems with these indicators and the way the choice reflected political and social concerns. Mutation, she argued, was defined as a "dangerous, threatening, or socially disturbing trait with implications for future human survival."[9] The studies based on these parameters were declared "inconclusive."[10] Scientists used the results to argue for further investigations into the genetic effects of radiation.[11] The survivors of the atomic bombings in Japan would remain a test population for a long series of new studies and approaches. Meanwhile, the publication of reassuring reports on the health of Japanese children served to pacify alarm over the deleterious effects of atomic radiation in the population who had suffered an atomic attack and also people back home who were contending with reports of atomic fallout from weapons testing.

Also a concern from the beginning and intensively studied were the somatic, cancer-inducing effects of the bomb. Little was known about the actual mechanisms by which radiation acted on organisms, but somatic effects (showing up during the lifetime of people exposed to radiation) and genetic effects (due to mutations in the reproductive cells and showing up in the next generation) were treated as separate effects.[12] This separation would eventually break down—not least because both effects could be studied at the level of chromosomes—contributing to a vastly expanded understanding of the genetic, including reproductive and somatic, effects of radiation and its connected risks.

Meanwhile, the decision by the American, British, French, and Russian governments to pursue the development of atomic energy for military and civilian uses was accompanied by vast new programs for radiobiological research. Radiobiological research centers were established in close proximity to nuclear energy research and development sites such as the Atomic Energy Research Establishment at Harwell in the United Kingdom and Oak Ridge National Laboratory in Tennessee, the uranium enrichment site of the Manhattan Project that later moved to civilian control.[13] At Harwell, the brief of the Radiobiological Research Unit, established in 1947 and funded by the Medical Research Council (MRC), was "to investigate the toxic actions of radioactive substances and to develop methods of protecting workers against them." This was before the fallout debate raised concerns about the effects of radiation not just on the workers handling radioactive materials but also on the population at large.[14]

For the MRC, the foremost government funding body for fundamental medical research in the United Kingdom, this was part of a broader commitment to harness the advancements of nuclear physics for biology and medicine and to advise the government on safety issues regarding the new field of nuclear radiation. This double commitment in many ways reached back to the role of the council in overseeing the development of radium therapy in the interwar years. It was substantially strengthened by the general advisory functions with respect to medical research matters the MRC assumed during World War II.[15] The MRC was directly responsible to Parliament (rather than to one of the ministries), and therefore regarded itself as independent from political pressures. It became responsible for much radiobiological work after 1945. As often noted, the situation was different in the United States, where the Atomic Energy Commission both promoted nuclear energy and funded much of the research assessing its health risk—a dual role for which it was often vehemently criticized.[16]

An important program, pursued both at Harwell and at Oak Ridge, was the long-term low-dose irradiation of vast populations of mice to establish safe limits for human radiation exposure.[17] Both centers profited from the on-site availability of nuclear reactors for their experiments. To establish mutation rates, researchers set up classic crossing experiments using the multiple recessive method, also known as single locus test. The first step involved developing a stock of mice that was homozygous for several recessive mutations that could be easily identified, thus allowing for quick scanning. Seven mutations were chosen, including characteristics such as brown coat, short ears, and pink eyes. In the experiments, sperm from irradiated wild-type male mice was used to fertilize females carrying two doses of a recessive mutant gene. If irradiation had produced mutation in the male, the offspring would show the mutation. If no mutation occurred, the offspring would look like wild type.

Despite large investments in the question of the long-term effects of low-dose radiation, the answer remained elusive. Irradiation experiments at Harwell and Oak Ridge—using millions of mice and other organisms as well as increasingly sophisticated irradiation regimes—continued well into the 1990s. Yet concerns about the genetic effects of radiation also stimulated parallel efforts to visualize mutations on the chromosomal level.[18] At Harwell, this aim was pursued in the Cytogenetics Section under Charles Ford. Ford became one of the key players in the establishment of human chromosome research in the 1950s, in the United Kingdom and internationally. His career, much like that of other chromosome researchers at the time, illustrates very well the changed opportunities for studying human chromosomes in the atomic age. The next section introduces Ford and some of the other protagonists of the following chapters, pointing to the multiple connections of their work to concerns surrounding radiation. Their career paths also demonstrate the close interactions between the select group of human chromosome researchers in the 1950s and early 1960s. The geographic proximity of various centers in the United Kingdom facilitated exchanges and this may well have contributed to the initial British dominance of the field.[19] Yet attention is also drawn to the participants from other countries who attended the first human chromosome standardization meeting in Denver in 1960. Participation at the conference was restricted to researchers who had already published a human karyotype showing forty-six chromosomes. This was a small club of thirteen people. Nuclear issues were never far from their endeavors.

Atomic Careers

A trained botanist, Ford moved to the newly established Radiobiological Research Unit at Harwell in 1949, after having spent three years at the Department of Atomic Energy at Chalk River in Canada, where he had studied the biological hazards of radiation using the root tips of the broad bean *Vicia faba*.[20] Once at Harwell, Ford and his collaborators set out to develop the technologies to study radiation-damaged chromosomes in mammalian cells—not without initial reluctance to put the elegant plant chromosome work aside. Plant root tips, with their rapidly dividing cells, were a convenient model system for radiation research, yet there was an urgent need to study the effects of radiation on chromosomes of mammalian organisms such as mice, rats, and rabbits, which could be more easily extrapolated to humans. Several technical advances in mammalian chromosome preparation techniques at the time were imported from botany. These included cell culture methods, as well as the use of colchicine, hypotonic medium, and squash techniques, all of which served to move away from working with embedded and sectioned tissue samples. Ford's contribution consisted in perfecting the squash technique and adapting the whole set of techniques to work with the most difficult of tissues, such as testis and bone marrow.[21]

Using his new techniques, Ford identified a specific radiation-induced mutation in mice based on an unequal translocation between two chromosomes that could be recognized easily under the microscope. At that time, the director of the unit, the hematologist John Loutit, was experimenting with bone marrow transfer, another line of research deeply intertwined with Cold War fears of atomic emergencies and the frantic search for an antidote to radiation damage in humans. Researchers at the University of Chicago had suggested that injecting bone marrow cells from a donor could protect mice from a lethal dose of radiation. They attributed the effect to a still-elusive hormonal factor, but Loutit postulated that the recovery might be linked to the repopulation of the host's bone marrow with donor cells. Ford recognized that his translocation, named T6, could serve as a suitable cell marker. By injecting the mice with T6-marked donor cells, the two researchers could show that all bone marrow cells in the surviving mice carried the mutation. The fact that foreign tissue could survive and grow in the donor pointed for the first time to the immune-repressive effect of radiation. It also showed the way to a possible therapeutic use of bone marrow transplantation and established the stem cell properties of bone marrow cells.[22] Ford decided to make the cytogenetic study of leukemic tumors a main line of his future research.

Meanwhile, his cytogenetic skills allowed Ford to be the first, in 1956, to confirm the new chromosome count that Tjio and Levan had suggested.[23] The Swedish group had made their observation on fetal lung tissue. In their paper, the researchers had stressed that "a renewed, careful control" of the chromosome number in human germ cells was necessary before their own results could be generalized.[24] Ford received a reprint of the paper by Levan, with whom he had corresponded before. As he later related to Levan, he was "so shaken by the possibility that man, after all, might have 46 chromosomes in somatic cells rather than 48" that he immediately got in touch with a surgeon who had offered him some human testis material before. A few weeks after that, he was satisfied that in all samples "there were 23 bivalents in normal first spermatocytes."[25] This was an important validation even if more counts were necessary to prove that the count held for all cells of the body and for all people around the globe.[26]

In the following years, Ford's cytogenetics laboratory in Harwell became a point of passage for other researchers entering the field. Several people who later made decisive contributions learned cytogenetic techniques from Ford.[27] His work also attracted the attention of clinicians, who sent him human tissue from patients under their care to be analyzed.

The other person in Britain who most energetically promoted human chromosome research and had the most expansive vision of its potentials was Michael Court Brown. Court Brown's career again emphasizes the increasingly close connection of radiation research, cancer research, and the study of chromosomes. Unlike Ford, though, Court Brown had a medical background and started his career as a radiologist. His clinical work stimulated an interest in research. To pursue this interest, he became a member of the scientific staff of the Medical Research Council in the Department of Medicine at the Postgraduate School in London. At first, he was interested in the acute effects of radiation, but soon he began studying the long-term effects of exposure to radiation. Reports from Japan indicated that survivors suffered from an increased incidence of leukemia. Court Brown, together with his colleague John D. Abbatt, provided evidence that a similar increase occurred in patients who had been treated with X-rays for ankylosing spondylitis, a rather common arthritic condition. Initial figures indicated a deeply concerning 50 percent increase. Court Brown hoped that the final figures would not appear "so alarmingly high and that statisticians may find some flaw in our figures."[28] The work caught the immediate attention of the MRC's secretary Harold Himsworth, and the first preliminary report on the experiments, published in the *Lancet*, went through various drafts and a long vetting process both in the MRC and by a select group of radiotherapists. Himsworth was particularly concerned that no "alarmist," not-fully-proven data trickle to the public and create a "scare about a new hazard of radiotherapy."[29] Beginning the report with a reference to the atomic bomb explosions and the study of the survivors, as Court Brown had done in his preliminary draft, seemed particularly "unwise," as it "might have suggested a prejudgment of the issue."[30] At the same time, a preliminary report recommended itself, "firstly, to warn the radiotherapists, certainly against repeated courses of treatment, and secondly, to stimulate people to produce information."[31] On Himsworth's urgent suggestion, Court Brown joined forces with the epidemiologist Richard Doll, who was already well known for his work on the link between smoking and lung cancer, to make the evidence "statistically water-tight" and "unassailable" and to follow up on the issue.[32] Specifically, the two researchers set out to determine whether they could draw up a dose-response relationship between radiation exposure and mortality from leukemia. Their monumental study followed up more than fourteen thousand patients who had been treated for the condition between 1935 and 1954 at eighty-two radiotherapy centers throughout the United Kingdom. Seventy-three radiotherapists and forty-five other colleagues helped extract and analyze the data, and the whole task required "military-like planning."[33] Doll considered it the "best-designed study" he had ever participated in and possibly his "best work."[34] A preliminary report appeared in the 1956 white paper *The Hazards to Man of Nuclear and Allied Radiations*, followed by a fuller report in the following year.[35] The data confirmed the dramatic rise in leukemia cases following radiation treatment and established that no safe threshold existed. Together with the survivor studies in Japan, summarized by Court Brown and Doll for the MRC report to Parliament,[36] this became the most important study to establish the carcinogenic effect of low doses of ionizing radiation.[37] The studies provided support to rising concerns about the long-term effects of radioactive fallout from the hydrogen bomb testing to which the Cold War powers were committed, although in the short run, exposure to radiation in the clinic was regarded as the more serious problem.

The leukemia study provided Court Brown with decisive experiences and contacts in scientific and political circles, setting the stage for his future career. Following the epidemiological study, he focused his attention on the new cytogenetic techniques that were just then being perfected. Influential in this respect were discussions he had with Ford at Harwell, where radiation physicists helped him calculate the radiation dose received by the spondylitis patients during X-ray treatment, at exactly the time when the new human chromosome count was being settled.[38] Court Brown's plan was to study the mechanism of the cancer-inducing effects of radiation. He also traveled to Japan to study the incidence of leukemia in the survivors and tried to convince the Atomic Bomb Casualty Commission to engage in a large-scale cytogenetic study (which was eventually conducted, as described later). By that time, Court Brown was heading the MRC Unit for Research on the Clinical Effects of Radiation at the Western General Hospital in Edinburgh. The unit later changed its name to MRC Clinical and Population Cytogenetics Unit, to better reflect the work the group was engaged in. The main remit of the unit was to achieve a fuller understanding of the development of cancer following exposure to radiation against the backdrop of rising concerns about the carcinogenic effects of nuclear fallout from atomic bomb testing and increased incidences of leukemia through radiation treatment.[39] Court Brown always mentioned this motivation in subsequent reports. His unit followed three lines of attack: epidemiological, cytogenetic (or subcellular), and clinical studies. For the cytogenetic part of the program, he hired Patricia Jacobs, a young geneticist from St. Andrews who he dispatched to Harwell to learn cytogenetic techniques from Ford. There, Ford and Jacobs, together with hematologist and bone marrow culture specialist Lazlo G. Lajtha, developed the bone marrow technique for chromosome analysis that formed the basis for much of the work on human chromosomes in the following few years.[40] Later, David Harnden, also from Harwell, joined the Edinburgh group.

The reference to the fallout context was not just rhetoric. In 1954 Court Brown, together with Tobias Carter, who ran the low-dose irradiation experiments at Harwell, was appointed joint technical secretary of the high-powered committee set up by the MRC to report to the British Parliament on the hazards to humans of nuclear and allied radiations. In his role, Court Brown was intimately involved with discussions on all aspects of radiation effects and with drafting the report. He himself submitted evidence in respect to the connection of X-ray treatment and leukemia, as mentioned earlier. The committee submitted a first report in 1956 and a second in 1960 to take into account further developments in the intervening years, including the publication of a United Nations report on the matter.[41] In preparation for the second report, Court Brown delivered a detailed list of all necessary amendments to the first report given new research.[42] Significantly, the second report, published just after the first wave of important findings in human chromosome research, contained an appendix that described the new tools and developments in human chromosome analysis to which radiobiologists like Ford and Court Brown had contributed so decisively. The reports framed the government response to radiation hazards and led to a reduction of the permissible limits for radiation exposure of nuclear workers as well as for the general population. In the MRC, work on the report stimulated a general review and expansion of its genetic program, with funds being made available to a variety of existing and new projects. An important part of those funds went into radiation-related genetic research at Harwell and Edinburgh.[43]

Also on the MRC committee was Lionel Penrose, by then the respected doyen of British human genetics and head of the Galton Laboratory in London. Although mainly involved with the study of mental disability, especially Down syndrome, Penrose was concerned with the overall mutation rate in humans. He shared the expectation that an increased mutation rate would also increase the incidence of known genetic diseases, including cases of mental illness that showed a genetic component.[44] This was regarded as "so costly to man and society" that scientists called for a reexamination of the permissible dose of radiation.[45]

Before joining the MRC committee, Penrose had chaired a subcommittee of the Genetical Society to consider the genetic consequences of atomic radiation in response to rising public concerns about atomic fallout following the first H-bomb explosions and their repercussions, as brought home by the heavy contamination of the Japanese fishermen and their catch on the *Lucky Dragon*, stationed outside the exclusion zone of the US Castle Bravo test on Bikini Island in March 1954. The aim was to publish "an authoritative factual report" that at the same time was constructive and emphasized the need for more research.[46] The topics the panel proposed tackling were the experimental production of radiation, with special reference to thresholds; the quantity of radiation produced artificially through military, industrial, and clinical activities; the relation between sterility and genetic damage due to irradiation; the effects of radiation in population genetics; and mutation in humans. In a memorandum "Genetic Mutation in Man" that Penrose drafted for the committee, he highlighted the need for an intense study of human chromosomes, especially in relation to "abnormal structure."[47] This was just before the recount of human chromosomes galvanized a more general interest in human chromosome research. In fact, Ursula Mittwoch, a recent PhD, was already pursuing some cytogenetic observations at the Galton.[48]

The subcommittee had only just started its work when the more highly powered MRC committee was announced, covering much of the same territory and including several of the same members. Some members of the society bemoaned the many "busy bodies" who suddenly showed an interest in radiation genetics.[49] Despite these misgivings, it did not seem useful to publish a separate report. Nevertheless, the subcommittee continued to function as a "clearinghouse" for information that it collected and passed on to the MRC.[50] Building on his draft paper for the subcommittee, Penrose's contribution to the MRC report consisted of three brief essays on the spontaneous mutation rate in humans and on the effects of spontaneous and induced mutations rates on single gene diseases and on mental diseases.[51] Penrose remained closely involved in human chromosome research, and although his own group, somewhat to his chagrin, missed identifying the chromosomal trisomy responsible for Down syndrome that he had studied so extensively, he was satisfied that chromosome research had put human genetics on a "very solid basis."[52]

Looking beyond the British scene to other participants in the first human chromosome standardization conference in Denver in 1960, we find similar career paths. Participants came from only a handful of countries, namely Britain, France, Sweden, the United States, and Japan. With the exception of Sweden, these were all countries committed to the development of nuclear energy for military and civilian uses, or, as in the case of Japan, deeply scarred by atomic exposure. Clearly missing in the round were Soviet representatives. Soviet researchers had made distinctive contributions to human cytogenetics in the 1930s, but this work came to an abrupt halt with the country's official support of Lysenkoism and the suppression of much genetic work.[53] Apparently, the Soviet Union was the only country among the many approached that did not publish the new chromosome nomenclature the Denver conference agreed on.[54] When cytogenetics resurfaced, it was in the context of biophysics and radiation biology.[55]

Besides Ford, who originally suggested the meeting, and Jacobs and Harnden from the Edinburgh unit, all from Britain, we find three researchers from Sweden, including Albert Levan, who—together with Tjio, by then based in America and also present—first suggested the new human chromosome count. Just like Ford, Levan had started his career as a botanist. Working at the plant-breeding station at Svalöf, where there existed a strong tradition in plant chromosome research, and using the root tip of the onion as model system, he tested the effect of different mutagens, including mustard gas and radiation, on chromosomes.[56] In the early 1950s, following a visit to Jack Schultz's laboratory at the Cancer Research Institute in Philadelphia, Levan started transferring his cytogenetic skills to study cancer induction in humans.[57] At the time of his common work with Tjio, he was heading the Cancer Chromosome Laboratory at the Institute of Genetics in Lund, which received funding from the then recently founded Swedish Cancer Society. The study of cancer chromosomes was a project shared by cytogeneticists like Ford, Court Brown, and others. Even though it strongly resonated with concerns of the atomic age, it is not clear how much these motivations weighted on Levan's decision to shift fields. In this respect, Levan is perhaps the exception to the rule, but his work nevertheless profited from heightened interest in cancer genetics.[58]

Jérôme Lejeune, the participant from France, who was invited on the strength of his work on the chromosomal basis of Down syndrome, had a medical background. At the laboratory of Raymond Turpin at the

Hôpital Trousseau, a pediatric hospital, he studied the mutational effect of radiation in the clinic by comparing the sex ratio of the offspring of irradiated male patients. The French Atomic Energy Commission (Commissariat à l'énergie atomique) and the French National Institute of Health funded the research. On the basis of their findings, the two researchers warned about the risk of the use of radiation in the clinic. The research was ongoing when Lejeune, together with laboratory intern Marthe Gautier, who had some previous experience with chromosome preparations, on Turpin's suggestion started the cytogenetic study of patients with Down syndrome that set him on a new path.[59]

On the other side of the Atlantic were Theodore Puck, the conference convener, whose work at the Biophysics Department in Denver included the study of radiation-induced chromosome damage in somatic cell cultures; Ernest Chu, working on cancer genetics at the Biological Division of Oak Ridge National Laboratory, funded by the Atomic Energy Commission; and Hungerford, from the Fox Chase Institute for Cancer Research in Philadelphia.[60] Together with Nowell, a pathologist with previous experience in radiation research from a stint as a US Navy medical officer, Hungerford, a trained cytogeneticist, had just published the first paper linking a specific chromosome mutation to a particular form of leukemia. Working with leukemic blood cultures, the two researchers also developed the peripheral blood method for chromosome analysis. The technique was based on the observation that phytohemagglutinin, a protein found in bean extract that was routinely used to remove red blood cells from blood preparations, also stimulated cell division in white blood cells. This made it possible to grow white blood cells (or leukocytes) in culture and make them amenable to chromosome analysis.[61] By offering a minimally intrusive method to gain human tissue for chromosome analysis, the peripheral blood method opened the way for karyotyping to be performed on a much larger scale.

Also present at the conference was Tao-Chiuh Hsu—credited with having published the first clearly spaced chromosome preparations based on the use of hypotonic medium (although still counting forty-eight chromosomes)—who was then equally working on chromosomes and cancer in his laboratory at M. D. Anderson Hospital in Houston. Sajiro Makino came from an established cytogenetic school in Sapporo, Japan, whose work gained new salience with the plight of the atomic bomb survivors.[62]

It is perhaps noteworthy that none of the American researchers mentioned here participated in the "genetic panel" of the National Academy of Sciences study on the biological effects of atomic radiation that published a report in 1956. The release of the report and its conclusions were tightly coordinated with the MRC report on the same issue, to which some British counterparts had so actively contributed.[63] Of course, these were the early days of human chromosome research, and the Denver conference itself is sometimes credited with having played a role in stimulating research in this field in North America.[64] Only a few years later, some of the participants at the Denver meeting would expose the genetic risks of the atomic age. Puck made the point, stating: "The current value of the maximum allowable dose was adopted in 1959. Since then we have become aware of a whole new group of human diseases which appear to be capable of being induced by radiation but whose importance and indeed very existence was unknown at the time the currently employed standards were adopted. These diseases constitute the genetic diseases due to chromosomal aberrations. . . . This set of diseases is so costly to man and to society that a re-examination of the permissible dose of radiation for large populations must be carried out as soon as possible."[65]

So pervasive was the connection of radiation research, cancer research, human chromosome research, and atomic concerns that it can seem even tautological to mention it, yet it was this connection that gave human chromosome research impetus, visibility, and urgency in the middle decades of the twentieth century. The increasingly tight connection of leukemia to radiation in the atomic age and its role in defining research agendas in cytogenetics as well as the use of chromosomes for radiation dosimetry illuminate these connections further. Together, they forged links between radiation, cancer, and chromosomes while also expanding the scope of genetic research to include somatic and reproductive transmission.

Chromosomes and Leukemia

That cancer cells showed abnormal chromosome pictures with multiple duplications, breakages, and rearrangements had been known since the work of Theodor Boveri at the turn of the nineteenth century. However, accurate description of the anomalies was difficult, and it remained unclear whether they were the cause or the effect of cancer, and therefore a secondary phenomenon. Access to suitable human material for chromosome studies remained an issue well into the second half of the twentieth century. The rapidly dividing cells of cancer tissue represented an exception to some extent. For this reason, observations were frequently made in cancer tissue (a fact that may help explain the vastly divergent human chromosome counts before the mid-1920s, when some kind of consensus on forty-eight was achieved). By the early 1950s cancer cells—before other human tissue cells—could be grown in a single layer on petri dishes, which opened up new experimental possibilities and provided, for instance, the basis for much of Levan's work on cancer chromosomes when he moved from plants to humans in the early 1950s (fig. 1.1).

By this time, increasing money and efforts were concentrated on the "dread disease" whose links with nuclear technologies and the concerns and opportunities of the atomic age became ever closer.[66] Great hope was placed in the use of new radiation therapies for cancer. Radioisotopes, produced by the same accelerators built to yield fissile material, were actively promoted as new therapeutic tools, thereby providing a peacetime shine to atomic science.[67] Soon hospitals received their own small accelerators for therapeutic uses. All these practices were greatly enhanced through the Atoms for Peace plan launched by Eisenhower in 1953, which created big business opportunities for nuclear technologies, including in the clinic.[68] Yet while nuclear technologies were heralded as new therapeutic interventions to treat cancer, radiation was also known to produce cancer. The association of radiation with leukemia, an invariably fatal disease at the time, was particularly strong, as it was often the first cancer to appear after exposure to radiation.[69]

Although studies on the genetic effects on the children of the survivors of the atomic bombings in Japan yielded no conclusive results, increased leukemia cases in the survivors were soon reported. The number of reported cases varied in accordance to the distance of the survivors from the epicenter of the explosions. Among survivors, leukemia became identified as the "atomic bomb disease."[70] Meanwhile, outside of Japan numerous studies documented increased leukemia cases in radiologists, in patient groups treated with X-rays, and in children whose mothers had undergone X-ray analysis. These results gained particular salience in view of worrying statistics that showed a steadily increasing rate of leukemia cases in the general population. On the backdrop

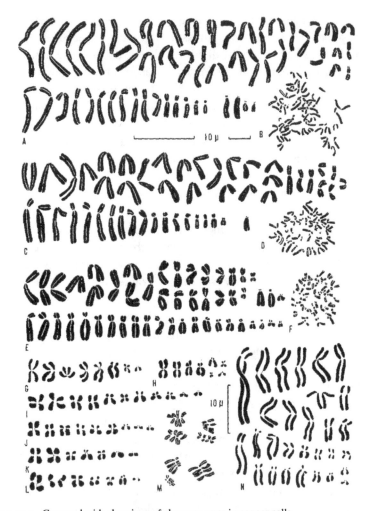

FIGURE 1.1. Camera lucida drawings of chromosomes in cancer cells.

Source: Levan, "Chromosome Studies on Some Human Tumors," 652, fig. 3. Reproduced with permission of Wiley Publishers.

of ever more intense atomic bomb testing by the nuclear powers and the proliferation of nuclear technologies, these observations focused the attention of researchers on the disease. Cytogeneticists took up the challenge. We have already seen how closely the development of cytogenetic techniques was connected to the study of leukemia. But cytogeneticists also hoped to contribute to the understanding of the causes of the disease.[71]

Initially, research focused on the study of bone marrow cells. Bone marrow tissue had to be extracted from patients in an invasive and painful intervention (usually by sternal puncture) before they could be grown in culture for cytogenetic analysis. Researchers had to stand close by to receive the fresh tissue. Jacobs, who was hired to begin the cytogenetic study of leukemia in Edinburgh, described how 50 percent of scientists watching the procedure "pass[ed] out cold." She belonged to the 50 percent that remained unaffected.[72] Once the peripheral blood technique became available, research conditions changed dramatically.

One of the first chromosome anomalies described after the establishment of the new chromosome count was the observation of an unusually small chromosome in the cells of patients with a particular form of anemia known as chronic myeloid leukemia.[73] The chromosome became known as Philadelphia chromosome, in line with a convention (otherwise rarely followed) that named unusual chromosomes according to the place where they were first observed. The Philadelphia group interacted closely with the Edinburgh group in confirming and interpreting the findings.[74] In particular, the Edinburgh group could confirm that the abnormality did not concern the Y chromosome as the Philadelphia group had first suggested and that the chromosome involved was not the same as the one that had just been shown to be present in an extra copy in Down syndrome patients, who also are diagnosed more frequently with leukemia.[75] The group also showed that the unusually small chromosome could be found in blood and marrow cells but not in skin cells, suggesting that it was a somatic rather than congenital condition. The work attracted much attention at the time, as it gave support to the thesis that cancer originated from one mutated cell that proliferated. It also proved the promise of chromosome studies for the field of cancer genetics, expanding the meaning of the term *genetics* to include horizontal (somatic) and vertical (through the germ line) transmission. Cancer development and hereditary diseases could all be studied with the same chromosome techniques.

Despite this initial euphoria and much subsequent work, Court Brown's assessment of progress in the field of cancer genetics at the end of the decade was rather disheartened:

Some nine years after [the discovery of the Philadelphia chromosome] we have not advanced one whit in our understanding of the significance of this chromosome, or in our attempts to answer the question whether its produc-

tion stands in a cause or an effect relationship to the development of the tumour. It is doubtful whether anything has transpired from all the great
amount of work, done in many centres over the last few years, on the chromosomal constitutions of human tumour cells of one kind or another which
leads us to be particularly hopeful that this approach will be rewarding in
furthering our knowledge of carcinogenesis.[76]

Looking for new approaches, the Edinburgh and other groups studied cancer viruses, their role in cell transformation, and their effects on
chromosomes. A few years later, chromosome-banding techniques offered new avenues for the study of cancer chromosomes, just in time for
human chromosome research to profit from the "war on cancer" campaigns launched by successive American presidents. As is well known,
nothing like the American cancer campaign, which played an essential
role in the formation and consolidation of the biomedical complex, ever
developed in Britain. Nonetheless, an MRC committee, set up in response to the American initiative to review how much money was spent
on cancer research in the United Kingdom, identified the study of chromosome structure and function as one of two areas (the other being tumor antigens) that looked most promising and therefore merited additional support.[77]

Banding was based on special stains that produced a characteristic striped pattern along each chromosome. The technique allowed researchers to distinguish single chromosomes more accurately and to
identify and locate small changes that had not been visible previously.
Using the new visual clues, Janet Rowley from the University of Chicago
was able to show that the small chromosome in patients with chronic
myeloid leukemia was in fact not a deletion but a translocation or an
exchange of material between two chromosomes (in this case between
chromosomes 22 and 9). The observation of translocations in other cancer cells followed, which set the stage for the study of specific genes and
their regulation in the etiology of cancer and for the design of new therapeutic interventions.[78] By this time, cancer research, including leukemia
research, had become a bandwagon, attracting researchers from many
different fields.[79] Yet the close association of atomic age concerns and
leukemia gave human chromosome research much of its original impetus and many of its tools while at the same time forging links between
cancer and chromosomes.

Radiation Dosimetry

With the postwar expansion of the use of atomic energy and the possibility of major accidents, as had occurred at Britain's new plutonium factory at Windscale in 1957 and the uranium-processing plant in Oak Ridge, Tennessee, the following year, nuclear protection services expressed an urgent need for a biological monitoring system that could supplement information from physical devices, such as film badges, which came into use in the early 1960s.[80] In contrast to the physical devices that merely recorded the received radiation dose, biological monitoring was expected to provide insight into the physiological effects of radiation. Such information was crucial in helping with the treatment and recovery of people exposed to radiation. Much effort was spent in trying to develop biochemical assays that would link the breakdown products of nucleic acids and proteins excreted in the urine to radiation doses. Yet such indicators proved unreliable because of the complexities of the human metabolism. Extensive experiments with plant chromosomes had already established dose-response relationships between different types of radiation and observable chromosome aberrations. The development of better protocols for the preparation of human chromosomes, and especially the peripheral blood technique, opened the way for human chromosomes to be used for the development of a radiodosimetric test. This project interested the Atomic Energy Authority in Britain as well as the National Radiological Protection Board (NRPB) located at Harwell, with much of the early work happening at the MRC unit in Edinburgh. Parallel investigations were also pursued at the Biological Division at Oak Ridge.[81]

In 1964 Court Brown wrote rather excitedly to the secretary of the MRC about developments in this direction: "I thought I would write and let you know that I think we are on the verge of considerable developments that could lead to a method of the biological control of low-dose radiation damage within about a decade. Basically the method is quite simple and is merely one of scoring the frequency of cells with chromosome aberrations in cells from blood cultured under standard conditions. The limitation on the method at present is that of the number of cells that can be processed by a human being without him or her becoming mental."[82] The work built on Court Brown's long-standing investigations on patients with ankylosing spondylitis in which he and his colleagues had been able to show chromosome aberrations following X-ray treatment. Yet more directly it regarded observations done on three workers accidentally exposed to radiation at one of the Atomic Energy Authority's establishments at Dounreay, on the north coast of Scotland. Chromosome aberrations could be grouped into two main categories: stable aberrations, such as balanced translocations, that led to viable daughter cells, and unstable aberrations, consisting of broken-off pieces, chromosomes with two restrictions, or ring chromosomes, which often led to loss of material during cell division and subsequent cell death (fig. 1.2). While chromosomes with stable aberrations were often difficult to distinguish from normal chromosomes, unstable aberrations were generally easier to spot and hence more suited for fast scoring under the microscope, as necessary for mass processing. To get significant results, many hundred cells in any single blood sample had to be assessed. For instance, one way of establishing radiation damage was to count the total number of rings and dicentric chromosomes per thousand cells per individual. For Court Brown, the practicability of the method depended critically on the development of automatic techniques for counting and analyzing chromosomes. He ended his first and many following letters on the issue by making a strong plea to the MRC for a crash program in that direction, adding strategically: "It goes without saying that we can adduce several other arguments, apart from that of radiation damage, to support the necessity for introducing automation."[83]

Himsworth showed himself "naturally" interested, but he urged Court Brown "to make a vow here and now to drop the word *damage* in relation to chromosome changes, and to discourage by every means in your power the use of this term by others in this context." He suggested the use of "some unemotive objective word like 'change'" instead. He explained: "If now you have got a technique which can pick up the effects of a dose of about 5 rads to the trunk from diagnostic radiation, the subject of the control of radiation is coming into a new dimension. And people are scared stiff of being damaged by radiation. Yet we know that thousands of people are exposed to these small doses and come to no harm. . . . After all, we are not absolutely certain of the significance of the changes observed."[84] Court Brown agreed to drop the word *damage* in relation to effects of low-dose radiation, but it does not seem that he managed to keep that promise. A few months later he reported to Himsworth about an unexpected, potentially explosive further development.

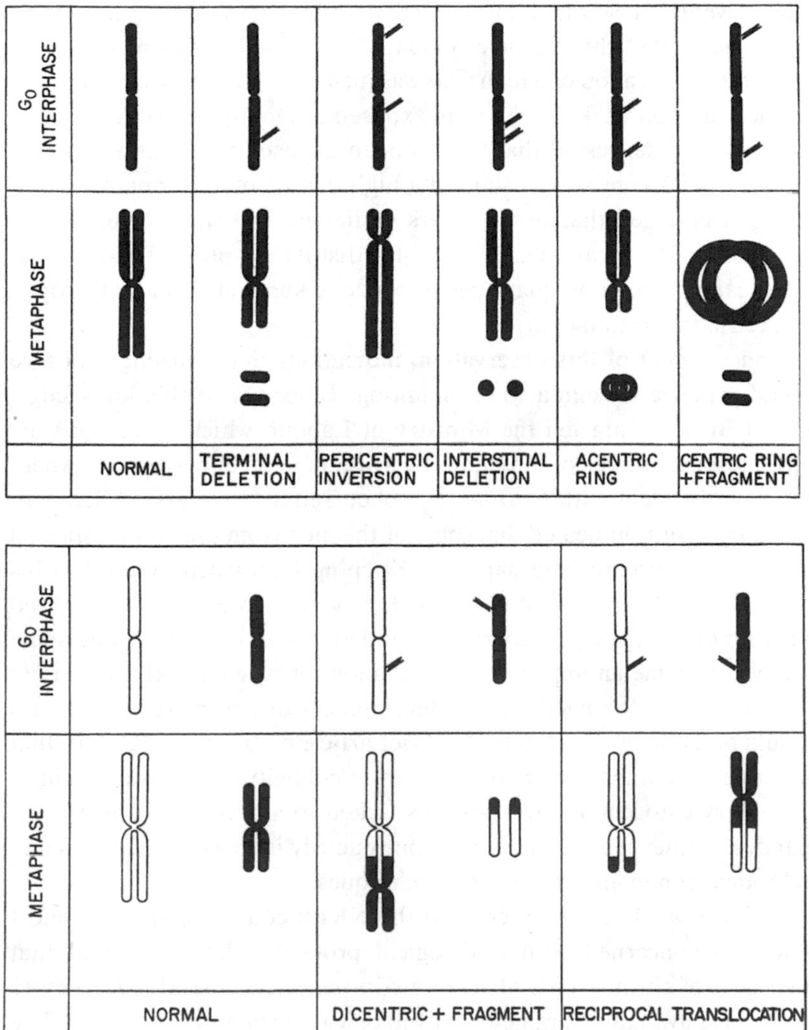

FIGURE I.2. Schematic representation of different types of chromosome aberrations: (A) in single chromosomes; (B) between chromosomes.

Source: Lloyd and Dolphin, "Radiation-Induced Chromosome Damage," 262, fig. 1, and 263, fig. 2. Copyright 1977. Reproduced with permission of BMJ Publishing Group Ltd.

There were indications that compounds like benzene were producing the same kind of chromosome "damage" as radiation. The work started from the observation of a man who was included as a control in the study of the radiation workers who were exposed at Dounreay. Nothing in the man's record suggested that he had been exposed to radiation, but his chromosomes consistently showed a high degree of abnormality. Eventually it emerged that twenty years earlier he had worked for several years as an electrical fitter in a coal-tar distillation plant. There he had been exposed to large quantities of benzene kept "at hand by the open bucketful for cleaning purposes."[85]

Independent of this observation, indications that benzene exposure could produce leukemia were mounting. Braced with this knowledge, Court Brown contacted the Ministry of Labour, which agreed to start chromosome studies of a number of workers in a rubber factory where an acute incident with benzene had occurred a few years earlier. Preliminary results indicated that some of the men from the factory showed "very gross chromosome damage." Keeping Himsworth abreast of his work, Court Brown pointed out that they could well have discovered "the tip of a very large iceberg." He added that if, indeed, chromosome analysis became an important tool for monitoring chemical exposure at the workplace, the need for the development of automated techniques would be even more pressing.[86] A brief article on the benzene work that he published in the *Lancet* received wide publicity.[87] The example indicates how cytogenetic investigations linked to atomic age concerns expanded in unexpected directions, continuously increasing the reach and relevance of human chromosome techniques.

A few years later, researchers at the NRPB could declare that "most scientists concerned with radiological protection have accepted that the yield of chromosome aberrations in peripheral blood lymphocytes is the best available parameter for biological dosimetry" (fig. 1.3).[88] The method and the NRPB monitoring service were originally set up in expectation of the need to assess radiation exposure in major radiation accidents. Given the fortunate absence of such events, the biological dosimetry service was used instead to check up on cases where film badges showed high or unexpected results or when workers failed to carry their badges. In contrast to film badges that monitored the cumulative exposure of ionizing radiation through the blackening of a photographic film, cytogenetic monitoring provided insight into the biological effects of radiation exposure, often before any other signs were visible. Chromosome

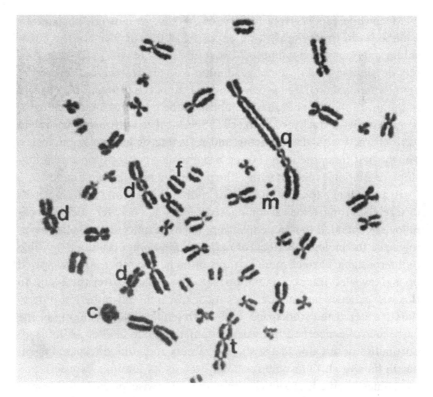

FIGURE 1.3. Photomicrograph of metaphase chromosomes in a human white blood cell showing various types of chromosome aberrations. The letters stand for the following: d = dicentric, t = tricentric, q = quadricentric, f = fragment (terminal deletion), m = minute (interstitial deletion), c = centric ring.

Source: Lloyd and Dolphin, "Radiation-Induced Chromosome Damage," 264, fig. 3. Copyright 1977. Reproduced with permission of BMJ Publishing Group Ltd.

aberration studies were also used to assess the radiobiological impact of various radiation treatments and the received dose of radiation in disputed clinical cases.

However, the relationship between the observed chromosome aberrations and their (longtime) biological effects remained more speculative. As Court Brown pointed out, there was "still no evidence one way or another" to interpret the chromosome aberrations as the initiating factor for the onset of leukemia. A possible interpretation that integrated cytogenetic observations and contemporary ideas about cancer viruses was that chromosome damage could produce cell clones that were unusually sensitive to cancer-inducing viruses.[89]

Meanwhile, reports from Japan showed that chromosome aberrations could be seen twenty years after exposure in the survivors of the Hiroshima and Nagasaki bombings.[90] These and other results in the then rapidly developing field of human radiation cytogenetics were discussed at an international conference convened at Edinburgh in 1966.[91] At the end of the 1970s, the Edinburgh unit published the results of a ten-year cytogenetic study of dockyard workers at the local Rosyth naval base who were charged with the servicing and refueling of nuclear submarines. For the first time, the data showed a significant dose-response relationship at radiation exposure within the maximum permissible limits.[92]

By providing a tool for monitoring radiation exposure, techniques for observing chromosomes accompanied radiation wherever nuclear technologies expanded to and people were exposed to its effects. Karyotyping made the biological effects of radiation (and other pollutants) visible in interpretable chromosome pictures often before other effects showed up and revealed traces of radiation exposure decades after the event. In this way genetics became part of cancer research and radiation protection. It entered the workspace, and—with global fallout peaking in the early 1960s—concerned the world population. The challenges and opportunities of the atomic age shaped careers and guided research agendas in human chromosome research even as its findings exposed, and sometimes pacified, anxieties about the otherwise often invisible and therefore so dreaded effects of nuclear technologies, with far-ranging implications for the policy arena. At the same time, human chromosome research developed a life of its own, detached from nuclear concerns. The next chapters deal with these developments and their implications. Chapter 2 starts by asking how human chromosomes found their place in the clinic far beyond the radiation context.

CHAPTER TWO

Chromosomes and the Clinic

The year 1959 was "the wonderful year" in human cytogenetics.[1] Building on the new consensus that humans have forty-six rather than forty-eight chromosomes, a string of papers reported unusual chromosome counts in patients with several known medical syndromes. In January of that year, three medical researchers based at the Hôpital Trousseau in Paris reported in a brief note in the *Comptes rendus* of the Academy of Sciences that three "mongol" boys they had tested carried forty-seven instead of the usual forty-six chromosomes. This was the very first time a specific karyotype was connected to a congenital condition.[2] Hardly a week later, researchers from the MRC Group for Research into the General Effects of Radiation at the Western General Hospital in Edinburgh published a paper in *Nature* suggesting that a male patient with feminized features, a condition also known as Klinefelter syndrome, had an XXY karyotype (instead of the expected XY karyotype for men or XX karyotype for women).[3] Besides providing another example of a chromosome anomaly, the findings challenged established views about the role of X and Y chromosomes in sex determination. In March and April of 1959, the Paris group followed up their original note with two additional reports on their observations of "mongol children" with forty-seven chromosomes.[4] Before the second of these reports, the *Lancet*—the leading British medical journal with a wide international readership—ran a set of three articles on chromosomal anomalies, flagged by an editorial that commented on the apparent revolution in course in human genetics.

The most novel of the three *Lancet* articles, based on a collaboration between researchers at Guy's Hospital in London and the Radiobiological Research Unit at Harwell, described the case of a young girl with gonadal dysgenesis, or Turner syndrome, who had forty-five chromosomes. She was missing one of the usual two X chromosomes. A second paper, emanating from another Harwell and London collaboration, reported the case of a patient showing both "mongolism" and Klinefelter syndrome and counting forty-eight chromosomes instead of the then-regarded-as-usual forty-six. Finally, an article by the Edinburgh group confirmed the additional chromosome in nine further cases of "mongolism" and identified it as a case of trisomy regarding one of the small somatic chromosomes (thus excluding the Y chromosome as a candidate). In September 1959, the Edinburgh group announced evidence for the existence of a "superfemale," or a woman with a triple-X chromosome.[5] Participants described this as a "wonderland of new discoveries." As the commentary in the *Lancet* remarked, "What next?" was the least-necessary question to be asked in this new field. There was an "enormous new territory awaiting exploitation."[6] In the following years, the *Lancet* would lend its pages to many more reports on chromosome anomalies, drawing clinicians' attention to the implications of the new findings. The commissioned essay "Chromosomes for Beginners," published in the journal in 1961, explicitly encouraged physicians to acquaint themselves with the new specialty and its techniques. The place to start, the essay suggested, was from "the exasperating pictures with which the pages of *The Lancet* have been so freely littered recently, and which are said to look like masses of squashed spiders."[7]

The possibility of tracing complex clinical syndromes to a change in shape or number of chromosomes that was detectable under the microscope very much impressed researchers and clinicians at the time. Penrose famously found "the photograph of the cell from the man with two extra chromosomes from which the intelligence level, the behavior and sexual character can be confidently predicted, just about as astonishing as a photograph of the back of the moon."[8] This was at a time when the first grainy images of the "far" side of the moon, taken by a Soviet satellite, had just appeared in the press. Hereditary considerations had a long tradition in medicine and had been central to the formulation of a biological concept of heredity in the latter part of the nineteenth century.[9] Nevertheless, chromosome anomalies were widely perceived as a completely "new group of human diseases" that opened the door for new diagnostic categories and clinical interventions.[10]

Historians who have taken up human chromosome research as a topic have made the clinical career of chromosome analysis and its role in the making of medical genetics the prevailing theme. They have considered the role of chromosome analysis, together with other technical, social, and institutional developments and constraints, in the construction of genetic disease categories; they have investigated the establishment of clinical cytogenetics and the contested shift of expertise from pediatricians to genetic specialists in the diagnosis of congenital diseases; above all, they have probed the social and emotional effects of chromosome analysis in the context of prenatal diagnosis and genetic counseling practices. In this context, the vexing question of persistent links with eugenic practices continues to loom large.[11]

The close and continuing association of human karyotyping with prenatal diagnosis provides support for the dominant clinical history of chromosome analysis. Yet as we have seen, the concerns and opportunities of the nuclear age (including its clinical ramifications but not restricted to them) provided much of the original stimulus for the postwar development of human chromosome research. These connections continued to play out in the clinical career of cytogenetics, not just in the field of cancer research. This was evident at the time. As the *Lancet* editorial on the first extraordinary crop of observations regarding chromosome anomalies from two key British laboratories pointed out, "It is not entirely accidentally that both the Harwell and Edinburgh laboratories are at least nominally concerned primarily with irradiation effects."[12] Yet despite the continuing debates that surrounded it, undoubtedly the clinical story provided a better public image for a field in formation than did the close association with concerns of the atomic age. These considerations aside, it is certainly true that the clinical story developed its own dynamic, with chromosome analysis not just making an impact on prenatal diagnosis and pediatrics but also holding promise for a broad range of fields, including cancer research, sex research, the study of mental disability, gerontology, and toxicology, to name a few. In an effort to map the expanding uses of chromosome analysis and the discussions that accompanied it, the early clinical career of the technique is an important chapter.

How, then, did chromosome analysis find its place in the clinic? To answer this question, we start with a closer analysis of the first descriptions of chromosomal diseases. In particular, the chapter shows how chromosome analysis inserted itself into ongoing discussions on the causes of mental disabilities and human sexual differences. Furthermore, it high-

lights the central role of clinical researchers, who functioned as mediators between the clinic and the research laboratory. Raymond Turpin in
Paris, Penrose and Paul Polani in London, and Court Brown and John
Strong in Edinburgh were instrumental in framing the questions and
providing the link between doctors and patients in clinical wards or special institutions and biological researchers who were mastering the latest
genetic techniques. At the same time, tissue samples, microscopic slides,
photomicrographs, expertise, personnel, skills, and tools moved between
the various settings where chromosomes were being established as a new
site of clinical interrogation.

Yet more was needed than the description of chromosome anomalies for chromosome analysis to become a medical technology and for
chromosomes, and genetics, to become entrenched in the clinic. To further probe this process, the chapter investigates contrasting projects pursued in hospitals and laboratories in Paris, London, and Edinburgh,
where several of the first chromosome syndromes were described, to ex-
ploit—or in some cases counteract—the new karyotyping technologies
while enrolling researchers, clinicians, actual and would-be patients, and
health services in the effort. It also explores some of the key tools, including surveys, registries, and screening services that were put in place
to mobilize chromosomes for clinical research and practice. Underpin-
ning all these efforts was the development of a standard nomenclature
that provided the basis for more detailed and comparable chromosomal
descriptions and diagnostic categories, including a standard visual representation of human chromosomes. Delving into the process of agreeing
on such standards draws our attention once more to the local, artisanal,
and visual practices that sustain work with chromosomes. Throughout,
the chapter highlights the complex and contested practices that turned
chromosome pictures into a diagnostic tool, encountered by an increasing number of people.

Chromosome Diseases

It was no accident that the first observed chromosome anomalies concerned mongolism or Down, Turner, and Klinefelter syndromes.[13] All
three conditions and, more generally, the genetic basis of mental disability and sex determination had been the subject of extensive clinical

observation, theoretical speculation, and biological experimentation. In the various attempts to answer the intricate questions posed by these complex phenomena, chromosome anomalies were implicated, hypothetically, early on. The specific chromosome anomalies that were eventually found were nonetheless surprising. They turned known syndromes into chromosome diseases and redirected clinical practice and research in unexpected ways. At the same time, human chromosomes and the techniques to visualize them entered new territories, raising expectations and sparking debates about their meaning and their uses.

The observation of chromosome anomalies in individuals with Down, Turner, and Klinefelter syndromes saw several of the same people involved, and the investigations were closely interlinked. It is nonetheless useful to start by following the three stories separately. When Penrose, in a letter to John Burdon Sanderson Haldane, his longtime colleague at the Galton who had recently left Britain to settle and work in India, tried to retrace the dense sequence of events that led to the "new knowledge on human chromosomes," his main message was that human genetics "now cannot be a one man show," but rather involved complex networks of researchers, clinicians, patients, samples, and technologies.[14] The concentration of researchers and clinicians in a rather small geographical radius, together with substantial funding for genetic research from the MRC, allowed investigators in Britain to establish the networks and exchange of tissue samples, technical skills, and knowledge that Penrose regarded as essential for developments in the field.[15] Yet other researchers also made decisive contributions to the new knowledge on chromosomes, and international exchanges and networks, including a series of international standardization conferences, became equally important features of the rapidly developing field.

The first suggestion that Down syndrome, then still known as "mongolism," was due to a chromosomal abnormality was in the early 1930s, when it was put forward nearly simultaneously by the Dutch eye doctor and geneticist Petrus Johannes Waardenburg and by the American geneticist Charles Davenport. Such abnormalities had been found in plants and animals as well as in human cancer cells, thus making it seem plausible that the phenomenon could also occur in humans. Davenport followed up on his suggestion by assisting the zoologist Theophilus Painter, a frequent visitor to his laboratory who, some years earlier, had helped establish the number of human chromosomes, "to get some perfectly fresh testicular material of a mongoloid dwarf" for testing. Painter could not find any irregularity, but Davenport did not give up on the idea that eventually some abnormalities would be discovered.[16]

Penrose, who had made his name with an in-depth clinical and genetic study of mental disability, also considered a chromosomal basis for Down syndrome.[17] He conjectured that individuals with the condition might have a triple (instead of the usual double) set of chromosomes. Several years later, he asked his colleague Ursula Mittwoch to test the hypothesis. She counted forty-seven or forty-eight chromosomes at a time when forty-eight was still considered the norm and definitely excluded a triploidy.[18] Following this result, Penrose apparently did not pursue the idea any further, although he continued to regard Down syndrome as one of the "most baffling" medical problems.[19]

A few years later, with the revision of the number of human chromosomes and the development of better protocols for chromosome preparations, the possibility that Down syndrome was based on a chromosome abnormality was once more put to the test. By 1958 at least four groups—one in Paris, one in Harwell (in collaboration with colleagues in London), one in Edinburgh, and one in Uppsala—were studying the chromosomes of individuals with Down syndrome, if with slightly different motivations.

In Paris, the initiative to study the karyotype of patients with Down syndrome came from Turpin, the head of the pediatric service at the Hôpital Trousseau and professor of therapeutics in the Medical Faculty. In the 1930s, Turpin, like Davenport and others, had pointed to the possibility that Down syndrome was based on a chromosomal anomaly ("une anomalie chromosomique"), brought about by environmental influences during the formation of the fetus. Turpin based his suggestion on an extensive clinical and familial study of Down syndrome births and cited the Bar eye mutation in the fruit fly, based on an "inversion" of a chromosomal region, as a model.[20] In the 1950s, Turpin, with the assistance of Lejeune, then a young medical researcher in his laboratory, continued his studies of patients with Down, investigating such aspects as sex ratio and palm prints, while running the consultation clinic at the hospital. Inspired by Tjio's display of his chromosome images, he encouraged Marthe Gautier, a medical resident (*chef de clinique*) in his laboratory with previous experience in cell-culturing techniques, and Lejeune to study the chromosomes of Down patients (fig. 2.1).[21]

At the MRC unit in Edinburgh, Jacobs was involved in a broad study

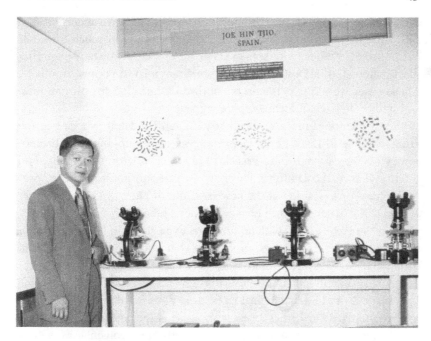

FIGURE 2.1. Tjio next to his chromosome display at the First International Congress of Human Genetics in Copenhagen in 1956.

Source: Aula Dei Experimental Station's Photographic Library. Reproduced with permission of Estación Experimental de Aula Dei (EEAD-CSIC), Zaragoza.

comparing the chromosomes of radiation- and non-radiation-induced leukemia. She started looking at chromosome preparations of Down patients, as they were known to be more susceptible to leukemia. Jacobs was also studying cases of Klinefelter syndrome. Some of the patients suffered from mental disability. This provided a further reason to expand the study to Down syndrome. The endocrinologist John Strong, who would soon join the MRC unit at Western General Hospital with the title of "honorary physician," contributed his clinical expertise and negotiated access to individuals with Down (then still mostly cared for in state-run institutions) by establishing contact with the local service responsible for mental disability. It should also be remembered that at this time most work on chromosomes was still performed on bone marrow cells that had to be extracted from the patient through sternal puncture. The chromosome analysis of a patient was thus anything but routine.[22]

At Guy's Hospital in London, Italian émigré Paul Polani had also become intrigued with the chromosomal theory of Down syndrome and supplied Harwell's Ford, with whom he was already in touch for other chromosome studies, with some samples. As a clinician he was particularly interested in Down syndrome children born to younger mothers. These cases were less common but showed a familial disposition, indicating that they might form a separate group.

Finally, Marco Fraccaro, together with Jan Lindsten, working at the Human Genetic Institute in Uppsala, was also studying the chromosomes of Down syndrome. Fraccaro, a young doctor from Italy, had spent a year at the Galton Laboratory in London before Penrose proposed him for a position at the new Institute of Human Genetics under Jan A. Böök, where he built up a cytogenetic laboratory.

That the Paris group, with no previous experience in human chromosome work, became the first to suggest that children with Down syndrome had forty-seven chromosomes took some of the British researchers by surprise. It speaks to the skills, especially of Gautier, in setting up a cell culture and cytological laboratory from scratch and with minimal resources.[23] The direct access to patients at the children clinic where Turpin had long been studying patients with "mongolism"—as well as the fast publication time of the *Comptes rendus* (submission to the right hands on Sunday could allow for a short report to be read at the academy's meeting on Monday and published a week later)—also favored the French group.[24] Following the preliminary note in *Comptes rendus*, the group published a second brief note on additional cases of the observed trisomy.[25] By that time, other groups were reporting similar observations. Emboldened, the French researchers presented "mongolism" as the first demonstrated "chromosome disease" (*maladie chromosomique*) and affirmed the opening of "chromosome pathology" as a new field of human genetics.[26]

The Edinburgh group was the first to confirm the Paris findings. Jacobs's more extended experience with karyotyping allowed her to conclude that the additional chromosome was a small somatic chromosome rather than a Y chromosome. Crucial help came from Penrose, who, with his unparalleled experience, was able to sort out the "mongols" from the "non mongols," which had confused Jacobs's counts. In a later interview Jacobs gave a vivid description of Penrose's visit to Edinburgh and the way he interacted with patients to assess their condition, often correcting the labels that had been attached to them in the institution that looked after them.[27] Surprisingly, but possibly in line with Penrose's more general attitude with respect to publications, his contribution was not mentioned in the published article. At this point, clinical judgment was still essential in sorting out Down cases from other mental disabilities and in guiding research, yet chromosome technologies would soon challenge this order of things.

Confirmation for the count of forty-seven chromosomes for individuals with Down syndrome also came from Uppsala, while Ford, analyzing a sample sent to him by Penrose, published the first case of a Klinefelter Down patient carrying two additional chromosomes.[28] It remained to Polani and Ford to unsettle the new consensus on the number of chromosomes in Down syndrome by publishing the first case of "a mongol girl with 46 chromosomes."[29] They suggested that the additional chromosome was attached to one of the larger chromosomes and therefore "hidden," a condition that was heritable through asymptomatic carriers of the translocation. Polani's decision to focus on the Down children of younger women had paid off, leading to the identification of a separate group of translocation Down. Other cases of Down patients with an apparent chromosome count of forty-six were found to be mosaics, that is, carrying some cells with forty-six and others with forty-seven chromosomes. This opened a completely new line of research. Together, the findings transformed what for a short time looked like a strikingly simple diagnostic test for a complex syndrome into a newly complex array of related but different conditions that could be distinguished only by looking at the chromosomes.

Like for the causes of mental disability, the mechanisms underlying sex determination had long attracted the attention of researchers and clinicians. The role of hormones, genes, and psychology were actively debated since the 1920s. The introduction of the Barr body or chromatin test in the 1940s, specifically directed the debate toward the role of chromosomes in sex determination. Named after its inventor, the Canadian anatomist Murray Barr, the test checked for the presence of so-called Barr bodies, small dark-stained round structures in the cell nucleus that were visible under the microscope. Barr and the PhD student Ewart Bertram had first noticed the bodies when, in an effort to investigate fatigue among pilots, they were studying the effects of prolonged activity on nerve cells in cats. The structure that was closely associated with the cell nucleolus, the large dark body at the center of the nucleus, showed up in some cells but not in others. Going back to their notes, they discovered that the structure was visible only in cells of female cats and absent in male cats. The same structure and the same sex correlation were found in nerve cells of other mammals, including humans, and in other tissues. The exact nature of the nuclear body remained unclear, but as it was present only in cells of females, Barr and Bertram suggested that it formed when two X chromosomes were present.[30] The researchers promoted the test as a technique to establish the sex of a person in cases where it was ambiguous, but it opened up as many questions as it managed to answer, and its interpretation stimulated a great deal of debate among cell biologists, geneticists, and clinicians. In the mid-1950s, the demonstration that the Barr body test, originally performed on skin biopsies, could be done on buccal smears—gathered by simply scraping some cells with a spatula from the inner lining of the cheek—brought the test within easy reach for many researchers.[31]

The Barr body test had dramatic consequences for individuals diagnosed with Turner or Klinefelter syndrome. Turner patients were viewed as girls who had failed to develop during puberty and showed a range of other physical characteristics, such as small stature and a webbed neck. In the Barr body test, though, they appeared as chromosomal males, leading to the suggestion that they be characterized more accurately as male hermaphrodites. Similarly, Klinefelter patients were regarded as males with underdeveloped sexual organs, including a series of other morphological and clinical features, but the Barr body test showed them to be genetic females.

Polani first became interested in the genetics of Turner syndrome when, in his ongoing study of congenital heart disease, he noted that some patients diagnosed with the condition showed a defect in the aorta that was much more common in men than in women.[32] The Barr body test seemed to support the idea that Turner patients were in fact males. Yet instead of taking the test at face value, Polani remained troubled by the "discrepancy between apparent and genetic sex" and started to entertain the unorthodox idea that Turner patients might have only one X chromosome.[33] The suggestion implied that the X and Y chromosomes played a different role in human sex differentiation than in the fruit fly, where unusual sex-chromosome distributions were found, but XO (i.e., the presence of a single X and no other sex chromosome) represented a male karyotype. At the time, what was true for the fly was very much taken to be true for humans. Not surprisingly, then, Polani's suggestion was dismissed as "fanciful and unacceptable" by the editor of the *Lancet* and as a "stupid idea" by Penrose, his teacher and mentor.[34] Yet Polani was not ready to give up. Enlisting the help of a colleague anatomist at Guy's Hospital who ran a cell culture laboratory, he first tried his hand at chromosome preparations to test his hypothesis directly. However, the technique did not prove "easy in execution and interpretation."[35] It should be noted that these attempts took place before Tjio and Levan's revision of the number of human chromosomes and before new protocols for chromosome preparations became more generally available.

Consulting with Penrose, he decided to revert to a classic genetic test instead by studying the occurrence of color blindness in Turner patients. Color blindness was an X-linked recessive condition that occurred more frequently in males than females. With the help of other clinicians, Polani managed to gather together twenty-five patients with what was, after all, a rare condition—to achieve statistically significant results. The findings were compatible with the presence of one X chromosome. This left open the question of whether a Y chromosome was present. By this time, Polani had made contact with Ford, who, on Polani's suggestion, had also participated at a conference on nuclear sex at King's College London. Apparently, the "plot" on that occasion was to get Ford more fully involved with human chromosomes.[36] Polani convinced him to perform the necessary chromosome analysis for him. The results confirmed his suspicion: the Turner patient analyzed had forty-five chromosomes, with an XO sex-chromosome complement. Given that the patient was "anatomically and psychologically" female, there seemed to be no reason to classify her as male. Rather, Polani and his coauthors suggested that the term *nuclear sexing* be dropped from the vocabulary and "more accurate if less striking terms" like "chromosome negativity" and "positivity" be introduced instead.[37]

Following these results, Polani was keen for Ford to also analyze the chromosomes of a Klinefelter patient.[38] Ford, together with Lajtha and Jacobs, then on her training course in Harwell, had already analyzed the chromosomes of a Klinefelter case in marrow cells supplied to them by the clinical pathologists William M. Davidson and David Robertson Smith, the creators of a variation of the Barr body test and conveners of the conference on nuclear sex at King's College Hospital Medical School in London. Submitting the marrow cells to their newly developed short-term cell-culturing technique that minimized the accumulation of in vitro artifacts, Ford and his colleagues had found a female karyotype, in apparent confirmation of the Barr body test.[39] However, the preparations had not been brilliant, to say the least, and there was scope to double-check the results.[40] Jacobs, by that time back in Edinburgh, received bone marrow cells of a Klinefelter patient from Strong, and somewhat to Ford's chagrin, they came out with their result first.[41] This time, the Klinefelter sample appeared as XXY. As an endocrinologist, Strong was undoubtedly aware of discussions around the chromosomal interpretation of the Barr body test and its implications for explaining sexual development, but unfortunately we know much less about his side of the story. In their joint article, Jacobs and Strong thanked the pathologist Bernard Lennox of the Department of Pathology at the Western Infirmary in Glasgow for performing the nuclear-sexing test. Lennox had moved to Glasgow from London, where he had assisted Polani in the nuclear-sexing test of patients with Turner syndrome. Yet in the following months, Neil MacLean, from the Pathology Department at Western General Hospital, would keenly take on this role, collaborating closely with the MRC unit on many future projects.

As the *Lancet* editorial of 1959 had predicted, the first chromosome anomalies were followed quickly by other striking observations. Among these, the association, in 1960, of an unusually small chromosome in white blood cells of patients with chronic myeloid leukemia raised particularly high hopes that chromosome research might provide new insights into the study of cancer (see chapter 1). The flood of case reports on chromosome anomalies in the *Lancet* was such that some readers complained, obliging the editor to start rejecting such manuscripts (fig. 2.2).[42] Around this time, cytogeneticists established their own journals, notably *Cytogenetics* (later renamed *Cytogenetics and Cell Genetics* and more recently *Cytogenetic and Genome Research*), which started publication in 1962 and became a leading journal in the field.

As the complex trajectories leading to the descriptions of Down, Turner, and Klinefelter syndromes as chromosomal anomalies in the late 1950s indicate, the techniques of human karyotyping developed by Tjio, Levan, and Ford inserted themselves into ongoing debates about the chromosomal basis of mental disabilities, human sexual differences, and the etiology of cancer. The Barr body test and its chromosomal interpretation in particular opened up a series of questions that seemed answerable only through direct inspection of chromosomes. The new techniques were eagerly picked up by researchers like Polani and Turpin, who had an intimate knowledge of clinical cases but also closely followed the research literature, including on the fruit fly. In many ways,

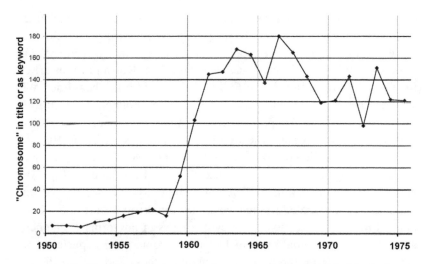

FIGURE 2.2. Number of articles reporting chromosome findings published in the *Lancet* between 1950 and 1975. Note the steep increase in articles in the late 1950s and early 1960s. The later decline reflects the existence of dedicated publications for chromosome research.

Source: Compiled by searching for articles where the search term *chromosome* occurs in title, abstract, or keyword using the Elsevier Science Direct Backfile online archive of the *Lancet*.

clinical researchers worked as mediators between the clinic and the laboratory. Clinical research received a boost in the postwar expansion of hospital-based medicine in the context of national health-care provisions in countries like Britain, France, and Sweden. This also allowed for the introduction of laboratory-based technologies like human karyotyping and medical specialties such as medical genetics into the clinic.[43] For their part, research workers like Ford and Jacobs, who themselves did not have a medical background, were keen to, and dependent on, entertaining close links with clinicians who were open to their endeavors and were prepared to provide access to patients in their care or human tissue samples from the operating room. This dependence created links that became productive in other ways. The new chromosome techniques did not require big apparatus, but they did require skills and expertise that took time and dedication to acquire. This led to a flow of human tissue, knowledge, skills, and personnel between the different sites, from which all sides profited in their own way.

News about the findings on chromosome diseases quickly moved beyond professional circles. In many cases the scientists themselves took up the pen to present their work and their implications to larger audiences by writing in popular science journals, newspapers, or magazines. For instance, Lejeune wrote an article on "mongolism" as a chromosome disease in the French popular science magazine *La Nature*.[44] The same topic was picked up by science writer Jean Rostand in an article for the Saturday edition of *Le Figaro litteraire* with the headline "Will We Soon Be Able to Heal Children Affected by Mongolism?"[45] In Italy, Fraccaro wrote several popular articles on the various chromosome findings, as did his landsman, the well-known geneticist Adriano Buzzati-Traverso.[46] According to Fraccaro, Buzzati-Traverso's articles, published in major media outlets like the weekly political magazine *L'Espresso*, made a "tremendous impact."[47] A few years on, guides for people who had been recommended to see a geneticist for chromosome analysis explained the "complex subject" of cytogenetics in lay terms.[48] The news about the chromosomal basis of Down syndrome raised special hopes for a new handle on the problem of mental illness and mental disability. Launching his special health initiative on this twin issue, in a 1962 awards ceremony in the White House, President Kennedy recognized Barr, Tjio, and Lejeune for their contributions. Among the other prize recipients was the Norwegian Ivar Asbjørn Følling, who in the 1930s had described phenylketonuria (PKU) as an inborn metabolic disorder. Left untreated, PKU led to mental disability in children. By the early 1960s, newborn screening for the condition and early dietary intervention had shown dramatic effects, inspiring hopes for cures for other mental conditions as well (fig. 2.3).[49]

Yet for karyotyping to become part of clinical practice, more was needed than just correlating chromosomal anomalies to specific syndromes. The step from chromosome analysis to clinical diagnosis and other types of clinical and public health interventions remained fraught with complications and was all but straightforward. The different avenues pursued by Lejeune in Paris, Polani in London, and Jacobs and others in Edinburgh are indicative of the various and contested ways in which karyotyping became a medical technology and chromosomes—and with them, genetics—entered the clinic.

Contrasting Trajectories

In Paris, Gautier, who felt that her work was not fully appreciated, left behind Turpin's laboratory and chromosomes to follow her career in

FIGURE 2.3. Prize recipients of the First International Award of the Joseph P. Kennedy Jr. Foundation in 1962. From left to right: Murray L. Barr, Samuel Kirk, John Fittinger, President John F. Kennedy, Ivar Asbjørn Følling, Jérôme Lejeune, Joe Hin Tjio.

Source: White House Photographs, John F. Kennedy Presidential Library and Museum Boston. Photograph by Abbie Rowe.

pediatric cardiology.[50] Turpin and Lejeune expanded their cytogenetic studies to include other conditions besides Down. In the following years, they contributed to the description of several new chromosomal syndromes. From the Paris unit also came new technical developments, including cell-culturing and later banding techniques. Throughout the 1960s, the laboratory under Lejeune's leadership was widely regarded as among the leading centers in the burgeoning new field of human cytogenetics. Lejeune himself took up public roles in connection with his expertise in cytogenetics. For instance, he sat on the expert panel that evaluated the first use of cytogenetic evidence in the courts and, on a French government grant, traveled to various countries to give courses in cell culture and karyotyping.[51] Yet increasingly Lejeune felt at odds with the direction in which the field was developing.

His unit profited from direct access to patients in the clinical wards. However, the particular clinical setting in which Lejeune worked, first at the Hôpital Trousseau and then at the Hospital for Sick Children, where he had followed Turpin after his appointment to head of pediatrics there, also meant that his chromosome work did not directly affect his clinical practice. At the Centre de progénèse, founded by Turpin, as well as in the consultation service for parents with Down children for which Lejeune was responsible, the focus was on the medical care of pregnant women and their newborns. In this context, Lejeune did not use the study of chromosomes to improve diagnostic categories or to inform counseling on genetic risk and reproductive choices. Rather, chromosomal diagnosis remained adjunct to clinical assessment, and Lejeune's all-absorbing aim became to develop a cure for Down syndrome. From the early 1960s he passionately advocated for a treatment of the condition based on a metabolic anomaly that he considered characteristic of the syndrome— presumably following the model of treatment for PKU.[52] He also investigated the factors responsible for non-disjunction of the chromosomes in development that triggered the trisomy, hoping to find ways to prevent the problem from occurring. He became strongly critical of early discussions on prenatal diagnosis, which had started to seem feasible in the mid-1960s with the development of amniocentesis and the ability to grow cells from amniotic fluid in culture.[53] This attitude brought him into conflict with other cytogeneticists who were at the forefront of such efforts. Things came to a head in the French debate on the legalization of abortion in the early 1970s, in which Lejeune took a strong public stance against it. Abortion conflicted not only with his strong Catholic beliefs but also with his medical commitment to treat, rather than discriminate against, individuals with Down syndrome.[54]

Among all the participants in the early chromosome observations in the London region, Polani was to make the biggest impact on the establishment of clinical cytogenetics and medical genetics more generally.[55] The contrast of Polani to Ford, working in relative isolation in Harwell, is instructive. Ford, in contributing what at the time was virtually unmatched expertise in mammalian chromosome preparations and in collaborating closely with clinically based researchers like Penrose, Polani, and Lajtha, had played a central role in establishing the new human chromosome count and later in many key early observations on human chromosome anomalies. For a few years, his laboratory was an "obligatory passage point" for work on human chromosomes.[56] Researchers and physicians from around the country sent samples to be analyzed or came to visit to learn the techniques themselves. Ford fielded a large volume of letters and accepted invitations to speak at meetings about the flood of results coming from his laboratory.[57] Yet once the initial excitement had settled, Ford reverted to applying his cytogenetic techniques, including several new technical developments, to the study of chromosomes in other animal species and to the exploration of more fundamental principles of cytogenetics. He continued this direction of work when, in the late 1960s, his group moved from Harwell to the Dunn School of Pathology in Oxford.[58]

In contrast, Polani set out to build on his cytogenetic incursions by hiring Ford's former junior colleague, John Hamerton, to his unit at Guy's Hospital to set up a chromosome laboratory there. With a generous endowment from the National Spastics Society he built up a Pediatric Research Unit at Guy's that combined cytogenetical, biochemical, immunological, developmental, and epidemiological research with clinical and genetic counseling services.[59] Polani had worked as pediatric research director for the society before. For the young society, whose main mission to that point was to care for individuals with cerebral palsy, the decision to provide substantial funds (an endowment of £2 million) for general research into congenital diseases was a bold and somewhat controversial step, but it reflected the rising expectation that genetic research could help solve medical problems. At the same time, the initiative highlights the important role of patient organizations in embracing and promoting certain lines of research, including chromosome research. Guy's Hospital, one of the oldest teaching hospitals in London and where Polani was already established as a researcher, offered a congenial institutional home for the expanding unit.

Polani's research focused on the study of mammalian female cells during meiosis (or germ cell division), with the aim of understanding the causes of chromosomal anomalies. However, early on he saw the need to integrate research with clinical service, which was delivered by the National Health Service. He orchestrated a first move in this direction by hiring John A. Fraser Roberts upon his retirement from the London Institute of Child Health at Great Ormond Street to continue his genetic counseling service at Guy's.[60] The unit was also at the forefront in offering a prenatal diagnostic service.

Polani's interest in this area might well have been stimulated by his collaboration, in 1959 and 1960, in a US-based research study group on "pregnancy wastage," initiated by the Obstetric Department at Johns Hopkins. The goal of the large-scale prospective study, involving forty thousand women and their (surviving and nonsurviving) offspring born in fourteen university hospitals around the country, was to identify the factors responsible for abnormal development that led to mental and physical disabilities. To this end, women were closely monitored throughout their pregnancy in a series of standardized clinical and biological investigations. The study was prompted by research into cerebral palsy, which explains Polani's participation. Yet it also followed on the heels of the thalidomide tragedy that saw thousands of babies born with congenital malformation of the limbs and other medical problems, and focused attention on fetal development.[61] Polani's participation was as an envoy of the European Headquarters of the World Health Organization (WHO). His special brief was to decide whether a parallel study should be started in Europe and more specifically in Ireland, the United Kingdom, and the Scandinavian countries—a plan that in the end was considered too costly and impractical.[62]

Cytogenetic studies were not part of the American study, but in his unit Polani initiated one of the first large-scale studies of chromosomal anomalies in miscarriages.[63] Miscarried fetuses were a "population cytogeneticist's dream" because of the sheer number of chromosomal anomalies that could be found.[64] As became clear, a large proportion of spontaneous abortions was due to chromosomal anomalies. Reported frequencies of chromosome anomalies in different test populations varied. This led to the possibility to study correlations between spontaneous abortion rates and exposures to environmental factors, such as drugs, radiation, and viruses, and parental age. The WHO published extensive guidelines to standardize data collection for comparative studies.[65]

From miscarried fetuses cytogeneticists turned to the chromosomal study of fetal cells that became available with the introduction of amniocentesis.[66] Polani's unit was at the forefront of these developments and in the later provision of a prenatal diagnostic service. It started offering prenatal diagnosis for chromosome anomalies in 1969—just two years after chromosomal analysis of fetal cells was first reported and two years after the passing of the Abortion Act in the United Kingdom expanded women's rights for legal abortion.[67] Other biochemical tests, most notably the measurement of alpha fetoprotein for the detection of neural-tube defects—a major research area in Polani's unit—were added later, with the diagnosis growing to forty different metabolic markers by the late 1970s.[68] In Polani's view, prenatal diagnosis allowed genetic counseling to move on from "crystal ball gazing" to concrete data on individual cases.[69] However, he also emphatically regarded prenatal diagnosis and selective abortions as only a "temporary expedient" until "primary prevention" of "inborn errors" became feasible. Research into the causes of cytogenetic anomalies—as Polani himself was pursuing with his studies of mammalian female cells during meiosis—was seen as a step in this direction.[70]

Polani actively sought the support of the National Health Service for providing a comprehensive genetic service, including testing and counseling, for the whole South East Thames region, which served as a model for other such services in Britain and internationally.[71] The unit also engaged in making the services more widely known, which was reflected in a steep rise in the number of cases referred. Data collected through the different screening services fed back into research by providing a continuous flow of material that could be mined for unusual cases and for epidemiological research, which found its way into scientific papers.[72]

With research and service closely integrated, tissue samples, chromosome preparations, photomicrographs, and karyograms could circulate relatively easily between the consultation room, the service laboratory and the bench. In the prenatal analysis of fetal cells, the cytogenetic test of necessity replaced any more comprehensive clinical analysis to which it could be related.[73] Although amniocentesis was always done together with ultrasound imaging, the chromosome pictures (or biochemical markers) rather than visual inspection of the fetus would clinch the diagnosis. This could sometimes be followed up when either the aborted fetus became available for analysis or the child was born.[74] In the consultation room the karyogram was shown to pregnant women and their partners to underline the point about an anomaly. It provided a "powerful visual diagnosis" that allowed the concerned party "to see"—even if perhaps not fully comprehend—the cellular manifestation of a genetic syndrome that was otherwise invisible.[75]

If in Polani's unit cytogenetics was put into the service of medical diagnosis and prenatal testing, the Edinburgh unit, under the directorship of Court Brown, was an early adopter of a population-based approach to the study of human chromosomes.[76] In Court Brown's vision, the relevance of chromosomes for medicine lay in the field of epidemiology. The unit, set up by the Medical Research Council in 1957 to study the clinical effects of radiation, was not just located on the site but also was closely integrated into Western General Hospital. A detailed agreement, based on the general principles laid down in the white paper on clinical research approved by the MRC, the Ministry of Health, and the Department of Health for Scotland in 1953, regulated the relations between hospital ward and research laboratory. One key request was that "clinical research cannot be dependent upon access to patients under the care of others. Senior clinical research workers must, therefore, have the full status of consultants or specialists."[77] Thus, Court Brown, a qualified radiologist, was appointed as consultant and put in charge of a ward—starting with six beds—in the radiotherapy department. He also had an outpatient and follow-up clinic geared toward the special research needs of his group. Consent rules for any patient intervention were in place. In addition, it was stipulated that "the primary concern in every case would be the effective and proper treatment of the patient."[78] As head of an independent clinical unit, Court Brown was also a member of the hospital's medical committee. The position of consultant gave him access to all other hospital services, including the Pathology Department, with which the unit collaborated closely, in particular with MacLean. The unit strengthened these links by appointing the endocrinologist Strong as honorary physician. Most researchers in the unit, including Jacobs, who was hired specifically to pursue the cytogenetic research program, did not have a medical degree, but they could rely on Court Brown for access to patients and medical expertise. Very quickly, cytogenetics—first of leukemia, then more broadly—became the driving research focus of the unit, and Court Brown had an ever-expanding vision of the areas in which chromosomes mattered. Many of the projects Court Brown initiated relied on clinical resources, and he actively sought the collaboration of medical personnel in and out of the hospital. Using these links, he started a set of infrastructure projects to support the unit's work.

The first descriptions of chromosome anomalies had only just been published when Court Brown presented the MRC with a proposal to set up a registry for abnormal chromosomes at the Edinburgh unit.[79] The idea behind this was to collect data on individuals with an abnormal karyotype and to compare their mortality patterns with those of individuals from the general population. More specifically, the aim was to study, first, the relation of karyotype abnormalities and cancer and, second, the relations of sex-chromosome complement, sexual phenotype, and cardiovascular disease. Yet very soon Court Brown saw a much-expanded use of the information collected in the registry.[80] The registry quickly grew into a central tool for population-based epidemiological studies in the unit. What is significant here is that the registry as envisioned by Court Brown relied on clinicians to submit cases for inclusion in the data set. In this way, the registry became an effective tool to recruit clinicians and physicians in the Edinburgh area to learn about the new karyotyping techniques and to collaborate on chromosome projects. The letter presenting the project and inviting participation that Court Brown sent out to fellow clinicians in pediatric, gynecological, and psychiatric departments, to general practitioners and physicians in mental institutions in Scotland and beyond, elicited positive to enthusiastic reactions. Respondents explained that they had been "fascinated by some of the recent discoveries made in the field of cytogenetics" and that they were "extremely interested in the whole question of chromosome counts." They ventured that there was "little doubt" that chromosome abnormalities would become "much larger a field than we are likely to suspect at present" and confirmed that they were "delighted" to participate in the project. One respondent hoped that Court Brown would make use of the "undoubted good will of the Association of Parents of Handicapped Children." Overall, about fifty clinicians expressed their interest in participating in the venture.[81] Asked to support the project after data collection had already started, the MRC fully backed the initiative even though it fell well outside the unit's original remit to study the general effects of radiation. Only one lone voice in the MRC found the whole project "premature."[82] Thus, the registry became a point of liaison between clinicians and researchers. Clinicians, most of whom did not yet have karyotyping facilities in their institutions, gained access to a new set of diagnostic tools while supplying researchers with cases of patients for inclusion in the registry.

Also in 1959, the unit set up a neonatal genetic-screening project. The first screening program, set up in collaboration with MacLean from the Pathology Department, was based on buccal smears followed by full karyotype analysis when an unusual picture was found.[83] The aim of the neonatal studies was to establish the frequency of abnormalities in the general population and thus provide a point of reference for other population studies, including the study of radiation-induced mutations that continued unabated. In addition, the newborn-screening program was meant to identify individuals who might need special attention and to provide data for genetic counseling. The long-term goal was to correlate anomalies in the children with parental age, social class, and ethnicity, and thus study the causes of the anomalies.

A new screening program based on full karyotype analyses was started in the mid-1960s. This second screening program was aimed specifically at identifying newborns with sex-chromosome anomalies for enrollment in increasingly contested prospective medical, psychological, and behavioral studies performed by clinically qualified personnel.[84]

The neonatal studies at Edinburgh were the largest of their kind conducted at the time, including tens of thousands of babies. They crucially depended on the close collaboration with the hospital ward. Jacobs has given a vivid description of the sampling involved, which relied on daily rounds in the neonatal wards: "These babies were lying in their cots. . . . And we just went in and pricked the heels. We just did it into a little tube wash and the babies didn't even wake up. . . . [We did this] every single day including Christmas day."[85] Early on, Court Brown, like Polani, saw the need to expand karyotyping to serve medical practice more generally. In 1960 he approached the Department of Health for Scotland on the matter. A couple of years later he drew up a memorandum on the current demands for cytogenetic studies and the requirements for a routine service to be forwarded to the Ministry of Health. On the basis of the number of routine requests directed at his own unit, he estimated a rate of requests of 150 per million population per year, a figure he expected to increase.[86] A few years later—and in view of what he depicted as the growing relevance of cytogenetics in such diverse fields as pediatrics, endocrinology, reproductive medicine, psychiatry, oncology, gerontology, criminology, and industrial toxicology—Court Brown called for the establishment of special facilities for cytogenetics within the framework of the National Health Service, backed by the "necessary advisory service from top class human cytogeneticists."[87] In Court Brown's view, the Edinburgh unit, which counted among the "three world centres of cytogenetics" (next to Paris and Philadelphia), could "naturally" serve as such a center, offering the whole range of cytogenetic techniques and the full expertise necessary to interpret complex cases. As Court Brown explained elsewhere, to fulfill such a function, a center had to have facilities for lymphocyte, fibroblast, and bone marrow cell cultures; for autoradiography; and for the study of both mitotic and meiotic chromosomes. It also had to have clinicians, pedigree researchers, and statisticians on staff, as well as access to computer facilities and the ability to consult with biochemical geneticists and blood group serologists.[88]

By offering to serve as the central reference center, the unit hoped to obtain information on additional cases for the registry. Through inclusion in the registry, single patients, whose chromosome variant might have been discovered through individual diagnosis in a clinical context, were enrolled in epidemiological studies that served to assess the public health dimensions of the new set of genetic conditions. Epidemiological studies based on chromosome surveys and the collection of data in the registry became the trademark of the Edinburgh unit.

Court Brown outlined the contributions of cytogenetics for clinical medicine and the need to provide cytogenetic service facilities for medical practice in a series of memoranda prepared for the MRC and in a well-received lecture before the Clinical Research Board at MRC headquarters. He also corresponded with the Ministry of Health on the matter.[89] As an indefatigable promoter of cytogenetics, Court Brown argued equally passionately for the relevance of the results and methods of chromosome research for radiation protection, the workplace, the courts, and the sports arena. Yet he never lost sight of the clinical dimensions. He also recognized that the clinical links provided a broad basis and strong public recognition for the emerging field of human cytogenetics.

Court Brown's sudden death in 1968 brought to an end some of his expansive projects that built on close collaboration with clinicians, although the unit continued its broad-based cytogenetic research program, spanning radiation and clinical research as well as the registry and neonatal screening program, well into the 1990s. The unit became the MRC Human Genetics Unit in 1988. Most recently, it merged with the University of Edinburgh Centre for Molecular Medicine and the Edinburgh Cancer Research Centre to form the new Institute of Genetics and Molecular Medicine, still (or again?) collaborating closely with clinical departments at Western General Hospital to gain access to key patient cohorts and clinical expertise.

With Lejeune, Polani, and Court Brown, we have followed different trajectories of chromosomes at the clinic. Lejeune's pursuit of a therapeutic agenda for chromosome diseases brought him into conflict with other cytogeneticists and the consequences of his own research on chromosomes, notably the establishment of prenatal diagnosis. The integration of cytogenetics into a broader agenda of medical genetics and the close integration of research and service turned Polani's unit into a model for developments elsewhere. The Edinburgh experience marked yet another trajectory, with chromosomes being enrolled in surveys, combined with other data in registries, and serving as an epidemiological tool that moved beyond the clinic and embraced variously defined populations as part of a larger strategy to expand the scope of chromosome research.

By the time chromosome diseases appeared as a new category, medical and clinical genetics departments already existed in several medical schools and hospitals in the United States, including at Johns Hopkins, the University of Washington, and the University of Wisconsin.[90] With time, these institutions also capitalized on the new developments, setting up cytogenetic laboratories, often with recruits from Europe. A leading example here is Victor McKusick, who, from the mid-1950s and with considerable institutional acumen, built up the thriving Medical Genetics Division at the Moore Clinic at Johns Hopkins Hospital in Baltimore. He embraced cytogenetics early on, hiring Malcolm Ferguson-Smith from Glasgow University in 1959—one of many postdoctoral researchers from Britain who would spend time in Baltimore—to build up a cytogenetics laboratory there. Trained as a pathologist, Ferguson-Smith had studied Klinefelter patients and, looking at testis biopsies, had become skeptical of the hypothesis that they were sex-reversed females, as suggested by the Barr body test.[91] Encouraged by Ford, who had just confirmed the new chromosome count reported by Tjio and Levan, he taught himself cell culture and chromosome techniques. Yet the cultures were poor, and he could not prove his point. The move to Baltimore offered him the opportunity to pursue the project further while building up facilities for chromosome analysis at the Moore Clinic. The beginnings, as often, were heroic—especially in hindsight. First assigned to a small cupboard off a secretarial office, Ferguson-Smith soon received permission to convert a nearby men's lavatory into a cytogenetics laboratory and a photographic darkroom. Grants from the US Public Health Service, and later the National Institutes of Health, provided funds for microscopes and other equipment as well as the hiring of research assistants. Ferguson-Smith himself had perfected his cytogenetic skills visiting Hsu and Levan (then on a sabbatical leave from Sweden) in Houston. Within a few months, "an active production line" was in place at the Moore Clinic to prepare, photograph, and analyze chromosomes. With the introduction in 1960 of the peripheral blood technique, the number of samples sent for analysis, both from the hospital and from farther afield, grew steadily. Thus, the "clinical cytogenetic diagnostic service [was] born."[92] Yet as in other cytogenetic units at the time, service and research remained closely aligned, as clinicians depended on researchers for the diagnoses, and researchers needed clinical cases and samples for their research.

In 1961, Ferguson-Smith returned to Glasgow, where, a few years later, besides running a research laboratory, he established a clinical diagnostic service at the newly opened Queen Mother's Maternity Hospital. The service, funded by the National Health Service, became one of the first to offer prenatal diagnosis in the United Kingdom, and—not unlike Polani's service in London—eventually it served the whole Glasgow area and beyond. Marie Ferguson-Smith, who had been doing much of the photographic work, was in charge of all cell cultures and also participated in research.[93]

For a while, McKusick had problems filling the vacancy in his laboratory left by the Ferguson-Smiths' departure. When he inquired with Court Brown in Edinburgh whether one of his experienced researchers would be interested in taking up the post at his clinic, he received a rather irritated rebuke.[94] The episode highlights the scarcity of senior cytogeneticists at the time. Eventually, Digamber Borgaonkar, who had been a research fellow in the laboratory with Ferguson-Smith, took up the post. Cytogenetic units were also set up in other departments at Johns Hopkins, notably under Barbara Migeon in the Department of Pediatrics, who had also been trained by Ferguson-Smith.[95] Cytogenetics was only one of several approaches pursued in the Moore Clinic, but McKusick recognized the scientific "glamor" that cytogenetics provided to medical genetics.[96] He himself made his mark as original author and chief editor of *Mendelian Inheritance in Man*, a continuously updated (eventually online) catalog of human genetic diseases and their genes, as well as an early and strong promoter of a full map of the human genome. Cytogenetics was only one tool among others that served these aims.[97]

Different practices sustained chromosomes and their analysis in clinics in Paris, London, Edinburgh, and Baltimore. Yet all enterprises built on a common ground: the standardization of the normal human karyotype and the creation of visual standards for work with chromosomes.

The Normal Human Karyotype

In April 1960, just over a year after the first descriptions of chromosome anomalies, several of the protagonists gathered in Denver for a three-day intense discussion on chromosome nomenclature. The object of discussion was the normal human karyotype, yet the need for conformity in the description of the normal karyotype was driven by the search for a standard against which to identify anomalies. What counted as normal and abnormal in the world of chromosomes was itself under construction.

Georges Canguilhem has drawn attention to the moment in Western history—epitomized by Claude Bernard's experimental physiology and its application to the clinic—when the pathological was viewed as an altered state of the normal rather than a different state altogether.[98] More recently, historians of science have linked the distinction between the normal and the pathological to specific, historically contingent practices of producing and communicating knowledge.[99] Thus, analyzing the processes through which standards and nomenclatures are agreed on can tell us much about the respective practices and objects at play. In the case of chromosome analysis, the establishment of standards was grounded in, and supposed to serve, hands-on visual practices of preparing, seeing, sorting, and describing subcellular structures under the microscope that underlay all work in the cytogenetic laboratory.[100]

Tjio and Levan were studying cancer chromosomes when they set out to investigate the normal human karyotype against which to compare the bewildering changes observed in the chromosomes of cancer cells. They were mainly concerned with the number of chromosomes, but they also made an effort to group and distinguish single chromosomes. In Levan's neat india-ink drawing in his 1956 article with Tjio, the chromosomes were subdivided into three groups—M, S, and T—according to the position of the centromere and the relation of the long arm to the short arm, or the arm ratio (fig. 2.4). This was a departure from earlier works in which chromosomes were lined up in one row from longest to shortest.[101] The authors recognized that the subdivision was "arbitrary, of course," as the arm relations varied continuously, but it was a first step in distinguishing single chromosomes and guided microscopic observation.[102] In later years, Levan defended hand drawing against the increasing practice of using photomicrographs in chromosome analysis on the grounds that, "in drawing, the worker cannot avoid focusing his attention in turn to every particular chromosome segment and thus to detect and estimate deviations from normality."[103]

By the time of his submission to the Denver conference, Levan proposed a subdivision of the forty-six human chromosomes into seven groups.[104] Others proposed similar subdivisions but with some variations. The groupings were guided by practical considerations. For instance, it was useful to make small groups with chromosomes that were difficult to distinguish visually.

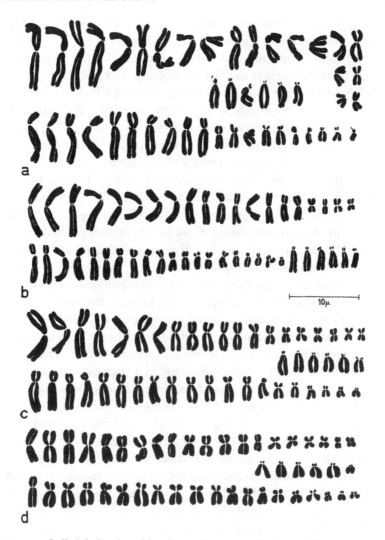

FIGURE 2.4. India-ink drawing of four human karyotypes (a–d), 1956. Note the arrangement of the chromosomes of each set into three groups (top, middle, and bottom row).

Source: Tjio and Levan, "Chromosome Number of Man," 4, fig. 2. Reproduced with permission of *Hereditas*.

The call for a standardization conference followed on the heels of the first string of observations of anomalous chromosome pictures. Agreement on a standard nomenclature was to facilitate communication between laboratories and thus bring order to the world of chromosomal disorders. The plan for the meeting is generally attributed to Ford.[105]

However, this does not explain why Theodore Puck hosted it in Denver. Puck, together with Tjio, whom he had invited to join him in Denver, had published an important paper confirming the new chromosome number, but by the time of the Denver meeting, Tjio had already moved on to Washington (fig. 2.5). In a biographical piece Puck gave the following account of the events leading to the meeting: "While ours was the first complete identification and classification system for the human chromosomes to be published, other papers appeared later proposing alternative numbering systems for chromosomes. This led to confusion in the literature. I corresponded with Charles Ford in England about what to do. He suggested that we attempt to straighten out the confusion by writing to all the people involved and suggesting a common human chromosome classification system. I decided, however, to call a conference to develop an effective system of classification of the human chromosomes."[106] As already mentioned, the decision was made to keep the meeting small and invite only cytogeneticists who had already published a human karyotype. This brought the number of participants to thirteen, twelve men and one woman. Puck introduced his colleague Arthur Robinson as an additional member to the group. Robinson, a pediatrician and later himself an accomplished medical geneticist, had recently joined Puck's laboratory in a transition to a career in research. He served as secretary of the group. Also invited were three "wise men" who were to act as arbiters if disputes occurred. They were David Catcheside from Birmingham, England; Hermann Muller from Indiana University; and Curt Stern from the University of California, Berkeley. All three were distinguished geneticists, but none had direct experience with human chromosome analysis. Puck thought this would make them impartial arbitrators (fig. 2.6). Puck applied for funding to the National Institutes of Health, but at the last minute the American Cancer Society stepped in. Puck credited the society with "having understood and supported the need for fundamental genetics in the coming era of medicine."[107]

The participants set their task as agreeing on a nomenclature that was simple, free of ambiguities, and flexible to accommodate future changes. The work accomplished by the group was condensed in three tables included in the rather brief final report. First, chromosomes—with the exception of the X and Y chromosomes, which kept their names—were serially numbered, as "nearly as possible" in descending order of length, "consistent with operational conveniences of identification by other

FIGURE 2.5. Cut-and-paste karyotype of a human female published by Tjio and Puck in 1958, before the Denver meeting. Possibly this (together with a male counterpart) was the first human karyotype constructed from single chromosomes cut out from an enlarged photomicrograph to appear in print. Note the numbering of chromosomes and their arrangement in different rows and groups.

Source: Tjio and Puck, "Somatic Chromosomes," 1232, fig. 4. Reproduced with kind permission of Jennifer Puck, daughter of Theodore Puck.

criteria."[108] The chromosomes were then subdivided into seven groups into which single chromosomes could be readily assigned following basic visual cues. The first table provided a "conspectus" of the groups. Each group was characterized by the serial number of the chromosomes it contained (e.g., group 1–3, group 4–5, group 6–12) and a brief verbal

FIGURE 2.6. The three impartial observers at the Denver standardization meeting. From left to right: David Catcheside, Hermann Muller (with camera), and Curt Stern.

Source: Curt Stern Papers, APSL. Reproduced with permission of the Library of the American Philosophical Society.

description of their main characteristics, including some additional practical hints for distinguishing them. For example, the description for group 4–5 read: "Large chromosomes with submedian centromere. The two chromosomes are difficult to distinguish, but chromosome 4 is slightly longer."[109] While visual categorizations guided the construction of the groups, these in turn were intended to facilitate microscopic observation and the visual distinction of chromosomes.

A second table displayed the actual work that had gone into defining the normal karyotype. It also shows the difficulty of the task the conference participants had set for themselves. In neatly arranged columns, the table listed the measurements of every single chromosome made from cells of "normal individuals" (except in one case) by the groups in Denver, Oak Ridge, Lund, Uppsala, Paris, and Edinburgh, all of which were represented by at least one participant at the meeting. Each chromosome was characterized by three parameters: relative length, calculated with respect to total length of all chromosomes contained in a normal haploid set with an X chromosome; the arm ratio, expressed as the length of the longer arm relative to the shorter arm; and the centromere index, indicating the relation of the length of the shorter arm to the length of the whole chromosome. A final column of the table gave the range for each measurement provided by the various groups. It revealed substantial variation. This was explained by the intrinsic difficulties in measuring small objects with fuzzy contours and by differences in preparation method, as it was noted that for every individual worker, measurements were more consistent.

Finally, a third table provided an overview about the correspondence of each chromosome in the new nomenclature with the numbering or naming proposed in previously published work. It made clear that all participants had to accept quite substantial changes to their own ordering schemes. Especially Lejeune, as representative of the French group, had to agree to rename every single chromosome, using Arabic numbers instead of a mixed system of letters and numbers. In his reminiscences about the meeting Hsu recorded his amazement at witnessing "the emotional involvement over minute details."[110] Yet at stake were not just questions of nomenclature but also questions of working practices that would have to change accordingly. Discussions were "often acrimonious" but apparently never to the point that the "wise men" had to intervene.[111] The final report acknowledged that the choice between different systems of nomenclature was "arbitrary," but that "uniformity for ease of reference is essential." For this reason "individual preferences" had been subordinated to the "common good."[112]

The study group also considered the usefulness of agreeing on a uniform way for presenting karyotypes or idiograms.[113] Yet in this case, "individual variation in taste" made "rigidity of design" seem "undesirable."[114] This decision may seem surprising given the importance of the visual display of chromosomes in analysis and in communication of results. Perhaps exactly for that reason, the matter was too contentious to fight out. The group nevertheless recommended that the chromosomes be arranged in numerical order, with the sex chromosomes near to but separated from the other chromosomes they resembled in shape and length. Similar chromosomes were to be grouped together and aligned at the centromeres, like being pegged to a clothesline. Despite consensus on these general rules, the final report as drafted by the study group and sent out for publication to journals around the world did not contain any visual representations of chromosomes. However, at least some people felt there was something missing.

Among the journals that reprinted the report with the proposed standard nomenclature was *Annals of Human Genetics*, published at the Galton Laboratory under Penrose's editorship. The actual report was prefaced by an editorial comment signed by Penrose. "For practical interest" it supplied its readers with a diagram of a set of chromosomes drawn using the means of the six sets of values published in the reports (fig. 2.7).[115] The idiogram followed most of the rules suggested by the report for pictorial representation, except for the fact that groups 1 and 2 were represented as one group and chromosomes were lined up at the bottom rather than the centromere. Penrose (who was not present in Denver) had gone through all the chromosome measurements submitted for inclusion in the report and found various arithmetic mistakes. These were also noted in the editorial.[116] Although Penrose was captivated by the visual evidence provided by the chromosome pictures, he consistently argued for the need for accurate measurements as a means to characterize and identify individual chromosomes and advance the field.[117] The chromosome diagram he provided combined the mathematical and pictorial elements that, in his view, sustained the work on chromosomes.

Researchers clearly welcomed the diagrammatic rendering of the new standard nomenclature. An enlarged version of the diagram, slightly corrected, titled "Average Measurements of Human Chromosomes—Denver System, Galton Laboratory 1960," with multiple pinholes in the corners clearly visible, was among the treasured papers of a former postdoctoral researcher in the laboratory.[118] The well-used chart had evidently served as a reference against which other measurements could be compared. The editorial comment containing the diagram was included

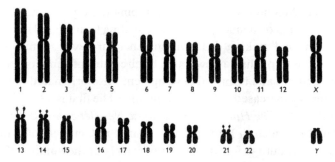

Means of the six sets of length values given in Table 2 using columns A and C only because values in columns B and C are sometimes inconsistent

	1	2	3	4	5	6	7	8	9	10	11	12	X
Short	41·5	31·6	30·8	16·6	16·0	20·6	18·6	15·3	16·2	13·8	15·0	12·4	19·9
Long	44·7	49·1	36·9	45·7	42·2	34·9	31·5	31·2	29·2	30·7	28·5	29·9	35·2

	13	14	15	16	17	18	19	20	21	22	Y
Short	4·8	5·2	4·8	12·1	9·0	6·4	9·6	9·3	4·2	3·9	2·0
Long	28·9	28·8	27·1	18·4	20·5	19·2	13·9	12·2	12·4	11·6	15·7

FIGURE 2.7. Diagram of human chromosomes drawn from average measurement values. Source: Penrose, "A Proposed Standard System," 319. Reproduced with permission of Wiley Publishers.

as an addendum to the report of the next standardization conference in Chicago in 1966, and so found even wider circulation.

The *Cerebral Palsy Bulletin*, published at Guy's Hospital in London, reproduced the Denver report in a special supplement but went one step further by including a photomicrograph of a normal male chromosome spread, followed by a karyotype with cut and paired chromosomes.[119] This time the layout of the group followed more closely the Denver convention, although again the clothesline was missing. At least one Denver participant strongly believed that including the images was a "mistake," as it seemed to "canonize" one particular way of ordering the chromosomes on the page.[120] Clearly, this was a contentious issue.

These squabbles notwithstanding, the Denver agreement was generally met with a "sense of relief" and accomplishment.[121] Even critics agreed that, although problems remained, the agreement was a step in the right direction.[122] The common nomenclature not only facilitated more effective communication across different settings. By defining the "normal karyotype," it also provided a basis against which to assess anomalies. With hindsight it became even clearer how timely this development was. At the Denver meeting, Hungerford first presented his method that allowed blood cultures to be used for chromosome analysis.[123] The technique would allow karyotyping to be performed in many more centers and on a much larger scale, leading to an abundance of new observations and making common nomenclature even more essential. The initiative to publish a newsletter to circulate information on chromosome findings and technical developments also originated at Denver, evidencing the increased sense of community. The first issue of the newsletter, titled *46—The Human Chromosome Newsletter* and compiled by Jacobs and Harnden in Edinburgh, appeared in October 1960 and was circulated to about 150 individuals. Two years later the number of copies sent out had grown to seven hundred, an indication of the increasing interest in human chromosome research.[124] Although standardization has often been viewed as a hallmark of modern science, it is instructive to note that a field not too distant from human karyotyping, blood group genetics, never achieved a standardized nomenclature.[125]

The Denver meeting set the pattern for later standardization meetings. In 1963 cytogeneticists convened for a Ciba Foundation guest symposium, "The Normal Human Karyotype," sponsored by the Association for the Aid of Crippled Children and hosted by Penrose in London. The aim was to assess how the Denver nomenclature had stood the test of time. The verdict was: surprisingly well. One change adopted at this meeting was that the chromosome groups were identified by the letters A–G rather than by the numbers of chromosomes they supposedly contained.[126] The change acknowledged the difficulty of assigning a definite number to each chromosome, as critics had pointed out. At a conference convened three years later during the Third International Congress of Human Genetics held at Chicago and sponsored by the National Foundation (originally the National Foundation for Infantile Paralysis, later renamed March of Dimes Birth Defects Foundation), the nomenclature of numerical and structural alterations, together with preoccupations about best practice in collecting and disseminating data, stood at the center of deliberations. Thirty-seven participants representing all major cytogenetic laboratories were present. At the time of the Denver conference, nobody would have foreseen "the wealth of variation which would soon be discovered," Penrose enthused in his introductory address. "Almost every day some new aberration is seen."[127] The introduction of computerized methods for various steps in the analysis of chromosomes further emphasized the need for a uniform and easily coded way to describe chromosome data. The suggested way to do this was a notation that indicated the total number of chromosomes, followed by a list of all the sex chromosomes present and the designation of the group or the chromosomes where an anomaly had been found. For instance, the notation for a girl with Turner syndrome would be 45,X and for a boy with Down it would be 47,XY,21+ (the plus sign indicating an additional chromosome was later moved in front of the chromosome number). More elaborate rules followed for the description of structural rearrangements, such as translocations, inversions, and fusions. This included choosing an abbreviation for the short and long arm of the chromosomes. To appease Lejeune, who objected to the fact that the nomenclature was all in English, the group settled on the letter *p* for *petit* (small) for the short arm. Following the next letter in the French and English alphabet alike, the long arm received the letter *q*.[128] The proposals of the Chicago conference also extended to the recording and storage of clinical and family data. Such data was considered "essential in all human cytogenetic studies," and the value of such studies depended on the completeness of the data collected.[129]

Complicating the distinction between normal and abnormal was the increasing awareness that not all chromosome anomalies corresponded to a (detectable) clinical symptom and that there was considerable chromosome variation in "phenotypically normal individuals." Such data emerged from the newborn-screening programs and other surveys. The assessment of the "normal range of variability" for various populations was seen as an important task, notably in view of determining the effect of radiation and other mutagens on the chromosomes.[130] Overall, the Chicago conference revealed the increasingly large apparatus building up around human karyotyping and its multiple relations to the clinic. It included new procedures such as the use of radioactive markers to distinguish chromosomes, an explosion of data, and a growing user community.

By the time of the Paris conference five years later, chromosome banding called for a vastly extended repertoire of rules for describing the human karyotype. In contrast to previous reports, the Paris report contained numerous representations of banded chromosomes, ranging from a series of cut-and-paste karyotypes to a complete chromosome-banding diagram.[131] With the verbal descriptions and notations necessary to describe chromosomes becoming ever more complex and unwieldy, visual representations that could be annotated in various ways gained new value. They corresponded more directly to the observational practice of cytogeneticists and allowed for direct visual comparison of banding patterns in the search for anomalies or in comparative studies of various species that attracted much attention at the time.[132] The diagrammatic representations of banded chromosomes published in the Paris conference report came to serve as increasingly taken for granted "visual standards" that guided research and structured communication while providing a common point of reference for a growing community of chromosome researchers.[133]

At the Paris conference the appointment of the Standing Committee for Chromosome Nomenclature was decided. One reason for this was that the number of workers in the field was now too big for a focused standardization meeting. The committee, chaired by Hamerton, would hence meet regularly. Later, the committee developed into an elected International Standing Committee on Human Cytogenetic Nomenclature. The first task of the new committee was to combine all available nomenclature recommendations into one international system for human cytogenetic nomenclature, and this has been updated and in circulation ever since.[134]

Although from the 1970s more attention was paid to representational aspects, it should be noted that the visual representation of karyotypes and idiograms was never completely standardized. Especially the spatial distribution of the groups (number of groups per line), the position of the X and Y chromosomes, and the alignment of the chromosomes (on the clothesline or the bottom line) continued to show variations.[135] Yet the karyotype as a genre certainly became iconic. Lined-up and ordered chromosomes served as supporting evidence in genetic consultation rooms, appeared in guidebooks for patients and doctors, and became a fixture in media reports on the new chromosome findings that often made the headlines. The X and Y chromosomes in particular became household names. Chromosome abnormalities like trisomies, monosomies, and translocations were conventionally highlighted with arrows in photomicrographs, karyograms, and idiograms (fig. 2.8).

When, in 1958, two of the key researchers involved in the new wave of work on human chromosomes reviewed the reasons for studying human chromosomes, they listed the establishment of ethnic differences, the etiology of cancers and leukemias, the identification of the "genetic sex" of intersex individuals, and a possible revival of comparative studies of animals and humans as questions to which karyotyping could make a contribution.[136] Only a few years later, human chromosome studies, backed by a standardized nomenclature, claimed growing relevance in a range of diagnostic categories and clinical services. If in the late 1950s,

FIGURE 2.8. Lejeune preparing a karyotype for a publication of the WHO.
Source: WHO Archives, WHO/12451, photo library reference WHO_A_008124. © World Health Organization/Paul Almasy, 1966. Reproduced with permission.

cytogenetics was still "a sleepy subspecialty of no interest to physicians," only a few years later the same discipline was hailed for having provided genetic medicine with "their organ" and a "sense of identity."[137] With the new organ came a new set of diseases, diagnostic categories, and tools to investigate the mechanisms that gave rise to unusual chromosome pictures and their effects. A growing apparatus of tissue collections, registries, diagnostic laboratories, and counseling services increased the space for chromosomes in the clinic, thereby spearheading the development of medical and clinical genetics. Implicit in this process was an expansion of the meaning of the term *genetic* in medicine: from relating to a hereditary condition running in families and traditionally studied by pedigree analysis it came to refer to all changes—hereditary, devel-

opmental and somatic—concerning chromosomes and genes.[138] Starting
from a handful of rare syndromes, the list of genetic conditions soon expanded, eventually leading to the expectation that all forms of disease
have a genetic component. In the words of one participant, cytogenetics not only provided an experimental foundation for human genetics but
also "convinced medicine itself that genetics had a place in the very foun-
dations of medicine, along with anatomy, physiology and biochemistry."
By doing so, "it expanded our understanding of disease, and even more
of health, . . . and created a new vision and new ways of prophylaxis."[139]
Treatment remained a more elusive aim but continued to guide research.
Penrose, for instance, greeted the description of Down syndrome as a
chromosome anomaly by declaring that the problem was "on the way
to a solution"—although the journey would still be long.[140] Practitioners
recognized that genetic technologies had social consequences and came
with a moral and financial price that ultimately had to be carried "by a
consenting public from whom the rapid advances demand quite sophisticated biological literacy and moral insight."[141] However, not always were
genetic explanations easily accepted. Nowhere did these issues come to
the fore more explicitly than in the polarizing debates raised by the suggested connection between the XYY karyotype and "criminal" behavior in the mid-1960s, which is discussed in the next chapter. To start with,
we need to consider the special status accorded to sex chromosomes and
their role in setting research agendas and defining identity, social behavior, and social policies.

CHAPTER THREE

X and Y

Following cytogeneticists in assigning a "personality" to each chromosome, we are perhaps justified in claiming that the X and Y chromosomes are by far the most famous of the lot.[1] They are the only ones carrying a letter for a name—albeit a letter that points to their unresolved nature. In ordered arrangements of the human karyotype they are set apart from the other chromosomes to which they correspond in size and shape, signaling their special status. People with a general education in biology may be unsure about the overall number of human chromosomes or whether every cell holds the same number, but they most likely know that females carry two X chromosomes and males an X and Y— even if the reality is more complex. The notation has become iconic and is used in scientific, medical, literary, artistic, and everyday contexts to define people and their sex (fig. 3.1). Several books have been dedicated exclusively to the X and Y chromosomes.[2] The X chromosome even has a poem written in its honor.[3] General cytogenetic treatises and textbooks regularly include separate chapters on sex chromosomes, and often sex chromosomes and their anomalies occupy a substantial amount of available space. For example, in his widely cited textbook on cytogenetics, John Hamerton, one of the early protagonists of human cytogenetics, dedicated a large part of one of two volumes to sex chromosomes and their pathologies.[4]

As this chapter presents, a specific constellation of technical, biomedical, and cultural reasons were responsible for the continuing focus on sex chromosomes. Following the X and Y chromosomes not only provides insights into a central set of questions and practices that propelled and shaped the study of human heredity in the middle decades of the twentieth century. It also leads to two of the most contested areas in

FIGURE 3.1. Female and male restroom signs.
Source: Photograph by the author (Berlin, 2015).

which human karyotyping played a role: first, the discussion of "criminal chromosomes" and its potential impact on criminal justice cases
and, second, the use of karyotyping in gender verification practices in
the competitive sports context of the Olympic Games. The first case has
become especially notorious in the history of human genetics and is often cited as an example of biased science.[5] The controversy started in
the mid-1960s, when a short paper in *Nature* reported an increased incidence of men with an XYY karyotype in high-security hospitals and
suggested a connection between Y chromosomes and aggressivity and
violence. This stretched well into the 1990s, when the last longitudinal
studies set up to clarify the issue ended. Here, the main aim is to ex-

plain the length and vehemence of the debate while also investigating its impact on the emerging field of human chromosome research.[6] Also on display are the role of the controversy in an incipient ethical discussion on clinical research and genetic screening in particular. The discussion leads us back to the cytogenetics research unit in Edinburgh where the controversial paper originated and where the longest-standing follow-up studies on the implications of the XYY karyotype took place. Chromosome testing for female athletes in the Olympic context was first introduced at the games in Mexico City in 1968 and became immediately controversial. Both cases serve to gauge the weight given to biology and human heredity to explain social behavior and inform social policies in the long 1960s.

"Pace Setters" of Research

At the turn of the nineteenth century cytologists trained their eyes on two chromosomes of unequal size that they observed in the cells of various insects. In a series of independent papers published between 1905 and 1912, Nettie Maria Stevens, working as a research and teaching associate at Bryn Mawr, a women's college in Pennsylvania, and Columbia zoologist Edmund B. Wilson, systematized these observations. They designated the unequal chromosomes as X and Y and identified them as sex chromosomes on the basis of their role in sex determination.[7] Even before then, the observation of the unusual chromosomes that could be followed through the formation of the germ cells and into the progeny provided evidence for the continuity and individuality of these cellular bodies.[8] For a long time, the X and Y chromosomes remained the only chromosomes that could be individually distinguished, and for this reason, they guided much research on chromosomes and their function.

The observation of the sex chromosomes in human cells remained difficult, and particularly the existence of the Y chromosome in human germ cells was contested. Counts for the human chromosome set varied widely, but in the 1910s the respected Belgian embryologist and cytologist Hans von Winiwarter suggested that females have forty-eight chromosomes (including two X chromosomes) and males forty-seven (showing an XO karyotype). In the mid-1920s, zoologist Theophilus Painter at the University of Texas at Austin based his effort to fix the number of human chromosomes for both sexes to forty-eight on his evidence for the existence of the small (Y) sex chromosome.

Painter had studied sex determination in insects and marsupials before getting hold of some testicular human material from so-called therapeutic castrations in a local mental hospital. Once he had convinced himself that humans, like insects, showed an XX-XY sex-chromosome pattern for females and males respectively, he argued for a count of forty-eight chromosomes in both men and women.[9] The count remained in place for over thirty years.

After his foray into mammalian chromosomes, Painter went back to study the chromosomes of insects. Together with Hermann. Muller, his colleague at Austin, he studied radiation-induced chromosome translocations and deletions in *Drosophila*. This work eventually led him to the description of the giant chromosomes in the salivary glands of the fruit fly and to the publication of an early map of the giant X chromosome, based on cytological observations rather than crossing-over rates.[10] Because X-linked recessive characteristics show up differently in males (who carry one copy of the genes) and in females (who carry two alleles of the same gene), the X chromosome has long remained a primary target for gene mapping in the fruit fly as well as in other organisms, including humans.

If initially work on human sex chromosomes lagged in comparison to chromosome studies in plants and insects, the situation had changed by midcentury when human sex chromosome research was setting the pace.[11] An important stimulus for the renewed interest in human sex chromosomes came from the development of the Barr body test in the late 1940s and early 1950s. Barr and his colleagues, who were responsible for developing the test, set the stage.

Realizing the potential importance of the test for "sexing" cells and tissues, Barr reoriented his laboratory. Joined by the PhD student and later collaborator Keith Moore, he focused his research effort on clarifying the relation between the "sex chromatin" and the postulated link with the X chromosomes. Barr also engaged with sex researchers, medical clinicians, geneticists, and individuals affected by unclear or contested sex assignment to explore the potential and limitations of the nuclear-sexing test. The term *sex chromatin* was introduced early on, although Barr and Moore soon suggested that the term *X-chromatin* was more "appropriate." As Moore explained, the term "omits the word 'sex' which sometimes has disturbing effects on patients with abnormalities of sex development, physical or psychological." In contrast to *sex chroma-*

tin, the term *X-chromatin* was "an informative term because it correctly indicates its X-chromosomal origin."[12]

In the mid-1950s, Moore, in collaboration with Barr, introduced the buccal smear technique for chromatin testing (the same procedure is still used for genomic testing today). Instead of working with skin or other biopsies that had to be extracted in intrusive procedures by qualified medical personnel, the new technique simply consisted of using a spatula to scrape off a few cells from the inner lining of the cheek. Transferred to a glass slide, fixed, and stained, the probe could be directly analyzed under the oil-immersion lens of a light microscope. For clinical diagnosis, about a hundred cells were examined, and the result was expressed in the percentage of cells carrying a Barr body (fig. 3.2). The quick and painless procedure was eagerly picked up by researchers and clinicians around the globe who were wrestling with a broad array of questions regarding sex assignment. Because of its simplicity, the test also lent itself to large-scale population surveys and screening programs, which began to be performed on selected populations, especially newborns, prisoners, and people in mental institutions.[13] As we saw in the preceding chapter, a conference on nuclear sexing convened at King's College Hospital Medical School in London, on the occasion of Barr's visit to Britain in 1957, proved an important occasion for bringing together people involved in these various endeavors and for forging new links between clinicians and human chromosome researchers.

The direct examination of the chromosomes was expected to answer many questions that the chromatin test had opened up, including the very nature of the Barr body. At the same time, the chromatin test propelled much of the early work on human sex chromosomes. The chromatin test became available when human karyotyping was still a time-

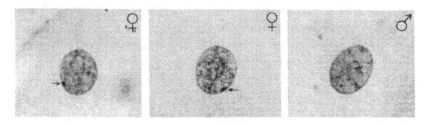

FIGURE 3.2. Barr body test on human buccal smears (two females, one male). Arrows point to Barr bodies in the preparations.

Source: Moore and Barr, "Smears," 57, figs. 1–3. Reproduced with permission of Elsevier.

consuming and laborious affair, most often performed on bone marrow tissue extracted by sternal puncture. Apart from the effort involved, researchers considered it unethical to ask healthy individuals to submit to the procedure, which was not devoid of risk.[14] Often a full karyotype would be performed only in cases when the chromatin test showed unclear or unexpected results. The questions raised by the test always concerned the sex chromosomes.

Clearly, the existence of the test alone does not explain the explosion of work on sex chromosomes. Rather, the chromatin test, as the work on sex chromosomes it initiated, found its place in an already-burgeoning field of sex research that saw the involvement of scientists, clinicians, and sex researchers of various stripes.[15] The research had immediate practical implications, as the eugenic movement as well as the feminist and gay rights movements were equally invested in knowledge of sex and ways to intervene in and regulate (or not) sexual processes. By the 1940s the chromosomal theory of sex was firmly established, but next to genes, hormones and psychological processes were considered to play an important role in the way sex and gender were articulated, thus defying any simple take on the issue.[16] Individuals who did not fit established gender categories were often caught in the middle of these scholarly debates, as made clear in an anguished letter penned by an anonymous correspondent signing with the pseudonym "Anomaly" and addressed to Barr.[17] Anomaly wrote to Barr in the hope that his test might prove that "sexual inversion" has a physiological rather than a psychological cause. Yet more often than not, the chromatin test did not bring the desired clarification. In the case of Anomaly, the test showed a male sex chromosome. How Anomaly might have received this information we do not know, as he died of an apparently unrelated heart attack before the result was available. In other cases the test certainly complicated matters. Most disturbing, perhaps, Klinefelter and Turner patients saw their sex reversed twice in the course of a few years.

The various cases later described as Klinefelter syndrome were grouped together and regarded as forming a distinct clinical picture only in the early 1940s.[18] The male patients associated with the syndrome had small testes and a variable set of other symptoms relating to incomplete male sexual development. The syndrome attracted little attention until, in 1956, several groups of researchers reported a positive chromatin test in Klinefelter patients, indicating that they were "genetic females." Researchers surmised that Klinefelter patients were in fact sex-reversed females or female pseudohermaphrodites. Follow-up studies showed that the cells of only some patients classified as Klinefelter showed a Barr body, leading to the distinction between "chromatin positive" and "chromatin negative" cases. A few years later, analysis of the chromosomes of Klinefelter patients vastly complicated the picture, dramatically changed entrenched ideas about the role of the X and Y chromosomes in sex determination, and finally reassigned individuals with Klinefelter a male sex.

A first published test reported an XX karyotype in a chromatin-positive Klinefelter patient, apparently confirming the interpretation suggested by the chromatin test.[19] Yet only a few months later, first researchers at Edinburgh and later other groups corrected the observation, announcing that Klinefelter patients had an XXY karyotype and describing the case as "human intersex."[20]

Interest in these results led to extended testing programs. Besides the most common XXY karyotype, researchers found a wide variety of other combinations of sex chromosomes in chromatin-positive Klinefelter patients, including up to four X chromosomes combined with up to two Ys (at least one Y was always present). Another striking result was the observation of cases of mosaicism, in which different cells in the body showed two and sometimes three different complements of sex chromosomes. Of these most often one was a usual XY male chromosome complement combined with an XXY one, but single cases also showed varying numbers of sex chromosomes in two or three distinct cell populations.[21] Researchers explained the presence of more than two sex chromosomes with the incomplete separation of the chromosomes in germ cell formation and mosaicism with non-disjunction of the chromosomes in early embryonic development.

Individuals with Turner syndrome experienced a similar double sex reassignment, showing up as "male" in the chromatin test but later being reclassified as female with an XO karyotype (see chapter 2). Seizing on these results, the testing of a young woman with early menopausal symptoms revealed an XXX karyotype. Somewhat paradoxically and apparently without much regard to the message sent to the patient, the case was referred to as "superfemale," in analogy to the corresponding case described for the fruit fly.[22] As in other early reports on chromosome anomalies, the case of the woman in her midthirties seeking medical help for her condition was described in some detail, providing information on her family history, the circumstances of her birth, and her

clinical history, complete with a photograph of her naked body in front
and side view, the eyes covered with a black strip in a conventional but
clumsy attempt to guard her anonymity.[23] Yet in the case description
there is enough detail to make today's reader ponder the medical odys-
sey of the woman and her personal experience of undergoing the test.

Following the new consensus on the normal number of human chromosomes, changes in chromosome numbers such as trisomies or monosomies were the easiest anomalies to find with the techniques available in the late 1950s. Theoretically, a whole array of different trisomies
could be expected. Yet only a handful of such variations, besides those
concerning the sex chromosomes, were found. These regarded trisomies
of chromosome 13, 18, and 21—all of which were described between 1959
and 1960. Of these, Down syndrome, which concerns the small chromosome 21 (more correctly even the smallest of all chromosomes), was
found to be the most frequent anomaly of all; the other two, known as
Patau and Edwards syndromes, are very rare and lead to severe and complex developmental problems.[24] As became clear, only trisomies of gene-
poor chromosomes—like the sex chromosomes and chromosomes 13, 18,
and 21—were viable. Many more trisomies (including the known ones)
were found in miscarried fetuses. Also this biological fact, then, contributed to a concentration on anomalies in sex chromosomes in the early
"heyday" of human chromosome research.

The early findings on sex-chromosome anomalies deeply unsettled
then-current ideas about the function of the X and Y chromosomes in
sex determination. Until 1959 the model for understanding sex determination in humans was the fruit fly, *Drosophila melanogaster.* As in humans, female flies have an XX karyotype and males an XY karyotype.
Calvin Bridges, studying flies with abnormal sex chromosomes, including XXY, XO, and XXX karyotypes, in Morgan's laboratory at Caltech
in the 1930s, had determined that the Y chromosome in flies was inert.
He postulated that male sex was determined by autosomal genes that in
females were counterbalanced by the two X chromosomes.[25] Researchers assumed that the same mechanism was in place in humans, but con-
tradicting this expectation, chromatin test notwithstanding, individuals with an XXY karyotype showed male characteristics and individuals
with an XO karyotype were females. The findings suggested that the
presence or absence of the Y chromosome rather than the relative dose
of X chromosomes to autosomes defined sex in humans. The discovery
of an XO female mouse in 1959 and an XXY male mouse a few years

later indicated that all mammals shared the same mechanism. The new mechanism of sex determination was perceived as a striking revelation at the time and added to the excitement about human chromosome research. Reviewing the crop of new chromosome findings in 1959, a *Lancet* editorial declared: "Perhaps most important of all . . . is the clear evidence that Y is after all no mere useless vestige."[26] The secretary of the MRC, which had funded much of the research, considered the finding of the sex-determining role of the Y chromosome "a fact of striking significance."[27]

Interestingly, the conclusions regarding the new role of the Y chromosome in sex determination were not drawn in the first paper on the XXY karyotype. Rather, Jacobs and Strong in their first communication on the case speculated on the mechanism that might have led to the presence of two X chromosomes and considered the possibility that the second X might in fact be "an autosome carrying feminizing genes."[28] Yet once the "masculinizing properties" of the Y chromosome were established, "the drive was on to find the [male] determinant," later identified in the sex-determining region Y (or the *SRY* gene).[29] Once considered of little biological importance, the presence, number, and length of the Y chromosome gained increasing importance in clinical diagnosis, in comparative studies of human populations, and as a marker of "maleness" in both scientific and popular discourse.[30] Reflecting on the experience of looking down at his own chromosomes in 1960, the British cytogeneticist David Harnden, perhaps the first person who had ever done this, recalled the "reassuring experience" that he had "a Y chromosome and only one."[31]

While initially the chromatin test had stimulated the chromosome analysis of the Klinefelter and Turner patients, subsequently these and other chromosomal findings helped clarify the nature of the Barr body. The first person to suggest that the condensed structure visible in the cells of various female mammals consisted of one (in the presence of a second) rather than two X chromosomes was Susumu Ohno, a young Japanese researcher working at the City of Hope Medical Center near Los Angeles. Ohno based his suggestion on meticulous observations he made under the microscope, where he followed the condensation of one of the two chromosomes present in a female cell through a complete cell cycle, first in the rat, then in a series of other mammals, and finally in humans.[32] His hypothesis predicted that there would be one Barr body less than the number of X chromosomes present in the cell. This better explained some of the chromatin test results found in individuals with sex-chromosome anomalies, especially the presence of two chromatin bodies in cells of women with a triple-X karyotype. Mary Lyon, a colleague of Ford at the Radiobiological Research Unit at Harwell, working on the genetics of radiation-induced mutations in mice, eventually formulated the hypothesis that one of the two X chromosomes in females is permanently inactivated, leading to the condensation seen in the Barr body test. Inactivation was considered to happen randomly, affecting one or the other chromosomes in different cells. Given the different origins of the two chromosomes—one stemming from the father, the other from the mother—this meant that women were genetic mosaics.[33] This hypothesis, together with the special mechanism of X-chromosome-linked inheritance, kept the X chromosome firmly in the center of further extensive studies directed at mapping the genes, understanding their role in development, and illuminating the mechanisms by which their functions were regulated. All the first genes mapped were on the X chromosome. By the time the first human gene—the Duffy blood group gene—was assigned to an autosomal (not sex) chromosome in 1968, the X chromosome map already counted sixty-eight loci. Among the genes found to be positioned on the X chromosome were important disease markers, such as those for muscular dystrophy, hemophilia, and glucose-6-phosphate dehydrogenase deficiency, the latter of which can lead to hemolytic anemia. Yet despite the attention lavished on the X chromosome, it was the Y chromosome and its newly acquired role as male determinant that first hit the headlines.

Chromosomes and Crime

On 25 December 1965, a brief paper with the title "Aggressive Behaviour, Mental Sub-Normality and the XYY Male" appeared in the journal *Nature*. It reported the results of a survey performed on inmates of the Scottish State Hospital (a hospital for mentally ill serious offenders) in Carstairs by researchers of the MRC Clinical Effects of Radiation Research Unit at Edinburgh in collaboration with hospital personnel. The survey showed a significantly increased incidence of individuals with an XYY karyotype among the inmates as compared to various randomly selected male populations. The authors interpreted this finding to

suggest that the additional Y chromosome, whose "masculinizing properties" had only just been established, predisposed its carrier to "unusually aggressive" and violent behavior.[34] In the words of its main author,
the paper "immediately caused a mayhem."[35] What was the issue?

The 1965 *Nature* article was not the first report of unusual chromosome findings in inmates of special mental and criminal institutions. Since
a connection had been established between chromosome anomalies and
mental disabilities in Klinefelter and Down syndromes in the late 1950s,
researchers were out to find more such links. Armed with the chromatin test, researchers in Canada, Sweden, France, and Britain performed
large-scale surveys of patients in mental hospitals. Following a Swedish study on 760 male patients in institutions for "criminal and 'hard-to-
manage' males of subnormal intelligence," M. D. Casey, a geneticist at
Sheffield University, surveyed 942 males in "two comparable institutions"
in England. He not only found an increased incidence of chromatin-
positive cases—twenty-one overall—but also studied their chromosomes.
He found seven who had an XXYY karyotype.[36] This result, which Casey
communicated to Jacobs before publication, contrasted sharply with comparative figures found in a survey of 2,607 "ordinary mentally subnormal males" performed in Edinburgh; there only two XXYY males were
found among the twenty-eight patients who had tested positive in the
chromatin test (overall the survey comprised the study of 4,514 males and
females in fifteen institutions, one in England and the rest in Scotland).[37]
The discrepancy made Jacobs and her colleagues at Edinburgh wonder
"whether an extra chromosome predisposes its carriers to unusually aggressive behavior."[38] It should be noted that this hypothesis included at
least two interpretive leaps: first, that the Y chromosome, recognized as
the "male" chromosome, was responsible for male aggression and, second, that an additional chromosome led to overperformance rather than
instability (as, for instance, in Down syndrome).[39]

Carstairs was in easy reach of Edinburgh and was among the institutions already included in previous surveys. Built as a mental hospital
in the late 1930s, Carstairs was first used as an army hospital before it
was returned to its original designation in the late 1940s. From the late
1950s, it was run as a maximum-security psychiatric hospital for patients
with a criminal record, the only institution of this kind in Scotland. The
hospital had two wings, one for "mentally subnormal" patients and the
other for "those with mental illness." Securing the collaboration of per-

sonnel at the institution, Jacobs and her colleagues obtained blood samples from 197 patients for karyotyping. Jacobs recalled:

> We were not allowed to see the patients, for reasons I simply don't know; the
> medical service for these patients was done by the local GP. He was a lovely
> man, and he would be in one room with the patients, while we were next door
> with our bottles. We would give him one and he would bring us back the bot-
> tle and a tiny piece of paper that had a coded number on it, the patient's date
> of birth and, for no reason I can think of, the patient's height. We didn't want
> the patient's height. We never asked for it, but it was there.[40]

In fact, in addition to height, weight, an estimate of the patients' intelligence, and reasons for admission were recorded.[41] Six patients declined
to participate, suggesting that a consent procedure was in place. All who
declined to cooperate were visited again after a few weeks. If they declined again, they were classified as "refusals."[42] In performing chromosome analysis on all participants, the Edinburgh researchers caught
anomalies like the XYY karyotype that remained undetected in surveys
that looked only at the chromosomes in cases that showed an unusual
chromatin test.

Little was known about the XYY karyotype. A first case was described in 1961 by a group of researchers in Buffalo, New York.[43] It
was found in a forty-four-year-old man who was tested because he had
a child with Down syndrome. The father showed no apparent symptoms, although further examination showed a complex reproductive
history. Of ten children the man had fathered with two different wives,
two died before birth and, in addition to the one child with Down syndrome, two others also showed developmental problems, suggesting
possible chromosome anomalies. Other children had been given up for
adoption anonymously and could not be chromosome tested. Overall
the results suggested a "possible familial predisposition to chromosomal
non-disjunction."[44]

A few other XYY cases were reported in the following years, but no
distinctive symptomatic picture emerged. The Carstairs survey for the
first time indicated that individuals with an XYY karyotype were on
average significantly taller than males with an XY karyotype. Instead
of the usual mean height of 170 centimeters, the Carstairs patients with
an XYY karyotype had a mean height of 186 centimeters. In the text—
unlike in the title of the article—the link with aggressive behavior re-

mained somewhat more hypothetical. As the authors stated in their con-
clusion: "At present it is not clear whether the increased frequency of
XYY males found in this institution is related to their aggressive behaviour or to their mental deficiency or to a combination of those factors.
We are attempting to elucidate this problem."[45] Despite Jacobs's recollections that her article and those that followed "immediately" caused
"mayhem," it is actually difficult to find direct evidence for this in the
surviving records. Indeed, in the following two years further studies
seemed to corroborate the initial findings. Casey and his colleagues,
studying a cohort of tall men in two maximum-security hospitals in England, where they had collected samples before, also found an unusually
high number of XYY men. While originally Casey had thought that the
increased frequency of XXYY males he and his colleagues had found
was "just chance," he now agreed with the Edinburgh researchers that
an extra Y chromosome had clearly "a part to play in increased stature
and antisocial behavior."[46] Meanwhile, a clinical follow-up study by William Price and Peter Whatmore, a psychiatrist and a pathologist associated with the Edinburgh unit, in collaboration with medical personnel
from the Carstairs hospital, added two further XYY cases to those previously reported. A preliminary comparison of their records with nine
control cases of the same institution did not suggest a difference in the
kind of crimes committed but confirmed their overall "aggressive and
violent" behavior.[47] An editorial in the same issue of the *Lancet* highlighted the "considerable psychiatric importance" of the XYY findings
as well as their "immense interest" from a "chromosomological" point
of view.[48] A letter to the editor a few months later expressed hope that
the studies would be known to all experts in prison affairs and criminal law.[49] An editorial in the *British Medical Journal* hailed the link
between XYY sex-chromosome complement and criminal behavior as
a major discovery and expressed the hope that it may lead to more appropriate treatment of the affected individuals.[50] Similarly, in an article titled "Genetics and Crime" published in the *Journal of the Royal
College of Physicians*, Court Brown, the highly respected head of the
Edinburgh unit, dubbed the finding of the XYY males "the most important discovery yet made in human cytogenetics" and potentially "a powerful lever to open up the study of human behavioral genetics."[51] Other
contributions in the special section dealt with societal, legal, and psychiatric aspects of crime and the criminal.

Somewhat more than a year after the original note in *Nature*, the continuing clinical studies of the XYY males of the hospital in Carstairs provided data that seemed to contradict at least partly the original findings. Going against the original claims, the studies indicated that XYY males were significantly less prone to violent crimes than the control cases chosen from their XY fellow inmates. This made the postulated link between the XYY karyotype and increased aggression seem untenable. Yet the comparison between the XYY males and the control group also indicated that XYY males, who all showed a "severely disturbed personality," were likely to get into trouble with the law at a much earlier age; that there was much less criminal background in their families; and that the usual corrective measures were largely ineffective. Price and Whatmore took this to confirm that the antisocial behavior of the cohort "was due to the extra Y chromosome," a congenital condition that was not transmissible, although they did not exclude the possibility that this was a "selected group" and that other individuals with the same chromosome set could show different features.[52] A full report of the cytogenetic and clinical studies of the inmates of the State Hospital in Carstairs, complete with detailed case descriptions of all the males identified as carrying an XYY karyotype, appeared in the following year.[53] The case reports presented a bleak picture of the life histories of the men that had been pieced together from family reports and institutional and penal records. They showed difficult-to-handle young children who underperformed at school and moved in and out of corrective institutions from a young age. Very often they were the "black sheep" in their families.

In December 1968 a comprehensive review titled "Males with an XYY Sex Chromosome Complement," compiled by Court Brown, shortly before his untimely death in that same month, appeared in the *Journal of Medical Genetics*. Carefully reviewing a growing body of literature on the XYY case, much of it from his own laboratory, Court Brown diplomatically hinted at some controversial aspects of the available studies and offered some careful qualifications.[54] The declared aim of the review was to gather all the information available to put together "a conspectus" of the "nature and identity" of the "XYY male."[55] Court Brown adopted a "historical approach" for his exposition. He distinguished three phases in the development of knowledge on the XYY man. The first phase was characterized by the apparently fortuitous accumulation of XYY cases. The second phase, initiated by the study of Jacobs and her colleagues, consisted of the systematic study of XYY males in maximum-security prisons and mental institutions, mostly in Britain. This phase revealed that there was also an increased frequency of women with an XXX karyotype in "hospitals for the mentally subnormal." Yet as Court Brown remarked laconically elsewhere, "nothing like the attention has been paid to them that has been given to abnormal males"—presumably because of the attention-grabbing XYY link with criminality.[56] The third phase, one that had only just started and was "likely to last for a long time," regarded the study of the "nature" of the XYY male.[57]

Discussing the first publication on the issue that had come out of his own unit, Court Brown, somewhat cryptically, considered that its finding "could have been unusual." Hence, looking at it retrospectively, it could be that "the reason for suggesting that 47,XYY males might be unusually frequent in maximum security hospitals was wrong."[58] This reassessment was based on the observation that the studies of Casey and others, which had prompted Jacobs's own study and on which her argument partly relied, did not in fact deal with groups of patients mainly defined by criminal activity but rather with groups of "hard-to-manage mentally subnormal" males who only occasionally got into trouble with law enforcement. If adjusted by IQ, these studies showed the same frequency of XYY individuals as in other mental institutions. Considering the psychological and psychiatric follow-up studies of the XYY males identified through the Carstairs survey, Court Brown also conceded that, "contrary to the impression given by the title of the preliminary paper by Jacobs et al. (1965) aggression was not an important feature of these men." He swiftly qualified this retraction by adding that it nonetheless appeared that "some XYY males may be extremely aggressive."[59] More generally, Court Brown was careful not to dismiss Jacobs's study as a whole. Indeed, in contrast to many of the following studies, it did not suffer from ascertainment bias by focusing only on tall individuals, which had skewed the frequencies in many later studies. Also, "whether right or wrong," the presumed link between the Y chromosome and criminality led to the survey of men in maximum-security hospitals and "to the build-up of knowledge about the XYY male."[60]

Looking ahead, Court Brown suggested that only newborn-screening programs that did not suffer from the ascertainment biases of studies of institutionalized populations and prospective longitudinal studies of infants found to carry the XYY karyotype could determine the developmental and behavioral implications of the extra Y chromosome. Assuming a frequency of one XYY karyotype per thousand male births, Court Brown calculated that there could be as many as eighteen hundred XYY males in Scotland, but the studies carried out to that point had revealed only about 2 percent of that expected number. This seemed to indicate that most XYY individuals were leading an unaffected life. Once the risk involved in developing "extreme behavioral aberrations" was known, routine screening would allow preventive measures such as surveillance and "adequate training" to be put in place at an early age when they were likely to be most effective.[61]

Court Brown's review article remained a point of reference for many years to come, and newborn-screening programs and longitudinal studies started in Edinburgh and several other centers around the world. Yet by the time the review appeared, news about the XYY male and his alleged criminal propensities had long left the realm of academic discussion and was hotly debated by expert legal panels, in the courts, and in the media, propelling chromosomes and their role in human behavior into the limelight.

Chromosomes on Trial

In 1968, Daniel Hugon, a thirty-three-year-old stable boy once in the service of the Aga Khan, was awaiting trial in Paris for the brutal murder of a prostitute in a hotel on the city's famed Place Pigalle three years earlier. After his suicide attempt, the court ordered a full medical examination, during which his chromosomes were tested. On whose suggestion this happened remains unclear, as no records relating to the decision have apparently been kept.[62] The test result showed an XYY karyotype. Confronted with this result, the court appointed an expert panel consisting of a psychiatrist, the cytogeneticist Lejeune, and a specialist in legal medicine to consider the scientific and legal implications of the XYY karyotype. The event was prominently reported in the international press, with the *New York Times* running a front-page article complete with a picture of Hugon's karyotype flanked by a photograph of the defendant (fig. 3.3).[63] At the trial in October, Hugon was found guilty but received a reduced sentence of seven years. At this point, the leading French daily *Le Monde* also dedicated a whole page to the case with two substantial articles, one by the paper's legal correspondent, reporting on the trial, and the other one, by the medical correspondent, providing detailed background information (complete with explanatory chromosome images) on the new genetic evidence presented to the court

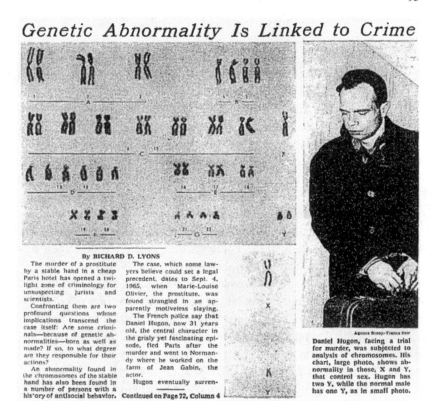

Genetic Abnormality Is Linked to Crime

By RICHARD D. LYONS

The murder of a prostitute by a stable hand in a cheap Paris hotel has opened a twilight zone of criminology for unsuspecting jurists and scientists.

Confronting them are two profound questions whose implications transcend the case itself: Are some criminals—because of genetic abnormalities—born as well as made? If so, to what degree are they responsible for their actions?

An abnormality found in the chromosomes of the stable hand has also been found in a number of persons with a history of antisocial behavior.

The case, which some lawyers believe could set a legal precedent, dates to Sept. 4, 1965, when Marie-Louise Olivier, the prostitute, was found strangled in an apparently motiveless slaying.

The French police say that Daniel Hugon, now 31 years old, the central character in the grisly yet fascinating episode, fled Paris after the murder and went to Normandy where he worked on the farm of Jean Gabin, the actor.

Hugon eventually surren-

Continued on Page 72, Column 4

Daniel Hugon, facing a trial for murder, was subjected to analysis of chromosomes. His chart, large photo, shows abnormality in those, X and Y, that control sex. Hugon has two Y, while the normal male has one Y, as in small photo.

FIGURE 3.3. *New York Times* front-page article on the trial of Daniel Hugon.
Source: Lyons, "Genetic Abnormality," 1. Reproduced with permission of the New York Times Company.

and the scientific, medical, and legal dilemmas it presented. As reported in the press, the expert panel agreed that "there is no born criminal" and that the social environment mattered greatly. Nevertheless, the carrier of an XYY karyotype had to be considered a "sick person" (*malade*).[64]

While the French expert committee was still deliberating, the American public learned that the Chicago mass murderer Richard Speck, who was on trial for the rape and brutal murder of eight nurses in one night, had been karyotyped. Newspapers (erroneously, as it later emerged) reported that Speck, too, had been found to carry an XYY chromosome set. Press reports about high-profile murder trials in Australia, Germany, and California in which cytogenetic evidence was presented followed suit. Despite various qualifications in these reports, the stereotype of the tall, violent, brutal, and not correctable XYY male, defined by his genetic makeup, started to take form (acne became an additional characteristic). Under such headlines as "Hidden Perils for Some Tall Men," "Genetics of Criminals," "The Sons of Cain," "Of Chromosomes and Crime," "Born Bad?," "Nature or Nurture?," "The XYY and the Criminal," and "The A-B-C of the X-Y Factor," dailies and weeklies in many countries presented the science of chromosomes and debated legal, penal, and philosophical questions of responsibility and free will, nature and culture, giving space to experts and scientists to voice their opinions.[65]

Flipping through the newspaper reports and the other articles appearing in the same pages, it becomes evident that, beyond the public interest in some of the most bizarre murder cases of the time, anxieties about public unrest and rising violence in the public sphere stood high on the agenda, especially in the United States, where the year 1968 was marked by student unrest, civil rights protests, anti–Vietnam War demonstrations, and the assassinations of Martin Luther King Jr. and Robert Kennedy.[66] These events increased the newsworthiness of a possible association of "bad genes" and aggression and propelled chromosome research into the limelight. Indeed, the chromosome studies found their place in a burgeoning field of research on the biology of aggressive behavior that was fueled by public interest in the matter. The topic stood high on the agenda of ethologists, psychologists, psychiatrists, and behavioral scientists, among others. The connection between brain function, brain disease, and aggressive behavior was under especially intensive investigation. Researchers explicitly defended the view that the problem of violence was not just a social but also (if not more so) a biological problem, and that new approaches to deal with the problem would come from biology as much as from social programs.[67]

Cytogenetic researchers would hardly have been surprised by or would have objected to the presentation of chromosome evidence in the courts. Well before the XYY karyotype became an issue, Court Brown, in a note in the *Lancet*, suggested that legal authorities should consider the findings of cytogenetics. He noted the fivefold increased frequency of males with sex-chromosome anomalies in "mental defective institutions" as compared to the newborn population and raised the question of the legal responsibility of people whose genotype might predispose them to delinquent behavior. He also considered the consequences of sex-chromosome anomalies for the legal status of certain marriages, as in the case of "phenotypic women" with a male karyotype.[68] In the context of the debate about the impact of the XYY karyotype, these considerations gained new valence and were picked up in various newspaper reports that provided background to the presentation of cytogenetic findings in the Hugon and Speck murder trials.[69]

Yet in their later recollections, researchers regularly complained about the way the media hyped up the reports on the XYY karyotype. For instance, in her speech accepting the William Allan Memorial Award of the American Society of Human Genetics in 1982, Jacobs expressed her anger about the "irresponsible attitude of the press": "There was the enormous amount of publicity given to our findings in the lay press. In common with many other individuals working in this area, I was quite unprepared for, and unable to adequately deal with, the media's sensationalist attitudes and blatant disregard for the facts. Unfortunately, this resulted in a great many people, both scientific and lay alike, receiving their first information about the possible association of an additional Y chromosome and antisocial behavior from these sensationalist and untrue accounts in the media."[70] Her critique echoed the one by Eric Engel, professor of medicine and chief of the Cytogenetics Division at Vanderbilt University's School of Medicine. Engel had been impressed by the first reports on the XYY individuals coming from Edinburgh. On hearing news of the "Chicago carnage," he wondered "whether the notorious suspect did not naturally fall into the YY class of 'supermales' described only months earlier in the British publications."[71] To further examine the problem, he wrote a letter to Speck's attorney, Mr. Gerald Getty, proposing a confidential chromosome test of his client. This was agreed and the test resulted negative. Quite independently, reports started to circulate in the press that six-foot-tall Speck was indeed a "YY super male." Engel recalled, "The persistent dissemination of this information, with such a flow of details, bated my breath."[72] Despite evidence to the contrary, the story was tenaciously kept alive.

There is, though, another side to the story. Having examined in detail the media reports on the "criminal chromosomes," science studies scholar Jeremy Green concluded that scientists contributed to the "false image" and the "myth" of the XYY male. According to Green, "The media have given wider circulation to this image, but it was not created by them."[73] Indeed, the idea of the link between XYY and criminality predated the trials and sensationalist media reports and made it possible for chromosome testing to enter the courtroom in the first place. Scientists also contributed to the media reporting. For instance, Lejeune—a member of the expert panel appointed to advise the court in the Hugon trial—is on record declaring that "everything about Hugon's life history . . . indicated that he was doomed to be a sick man from the moment of birth and that his hereditary affliction prevented him from exercising normal responsibility."[74] This statement fit neatly with his plea to lower the penalty for Hugon, but it also depicted a bleak picture of the genetic predicament of individuals with an XYY karyotype. Only one year later, accepting the William Allan Memorial Award of the American Society of Human Genetics, Lejeune argued passionately against a new cytogenetic-based eugenics that might rid society of XYY and other individuals with anomalous chromosomes.[75]

Green also cited the biologist Mary Tefler from the Elwyn Institute in Philadelphia, the author of a cytogenetic study in local prisons, who told a *New York Times* journalist that she could identify Speck as an XYY man just from his picture in the press.[76] Scientists did not just rely on journalists to intervene in the public debate but often took up the pen themselves. The British-American anthropologist and writer Ashley Montagu, who had been a rapporteur on the UNESCO Statement on Race and had written extensively against race as a biological concept, contributed a piece titled "Chromosomes and Crime" to the popular magazine *Psychology Today* that appeared in the midst of the Hugon and Speck trials. The article carried as a frontispiece a dramatic picture of an anguished young man whose shirt was pulled back to reveal a large tattoo lettered across his chest that read "Born to raise hell," as Speck apparently sported (fig. 3.4).[77] Consistent with his general position, Montagu argued that "genes do not determine anything." Yet there seemed to be exceptions: "Some individuals, however, seem to be driven to their aggressive behavior as if they were possessed by a demon. The demon, it would seem, lies in the peculiar nature of the double-Y chromosome complement."[78] Montagu went on to acknowledge that there was a "wide spectrum of behavioral possibilities from totally normal to persistent antisocial behavior" for the XYY individual. Nevertheless, society had to consider how to deal with such individuals. Montagu advocated for the chromosome testing of all infants at birth and "a program of social therapy" that would prevent children with an XYY karyotype from developing antisocial behavior.

Later correspondence adds a twist to the story. Apparently, the dramatic picture that accompanied the article was added by the publisher, without Montagu's knowledge. The picture, said Montagu, suggested the

FIGURE 3.4. Photograph accompanying Ashley Montagu's article "Chromosomes and Crime" in *Psychology Today* (1968). The photograph, credited to staff photographer John Oldenkamp, became an issue of contention between the author and the publisher.
Source: Montagu, "Chromosomes and Crime," 42.

opposite of what the article did and sent the wrong message. When the illustration was used in a reprint of the article, he threatened to sue the journal for damaging his reputation, but in the end he refrained from such action, given the wide success of the piece with lawyers, students, and teachers.[79]

According to Green, entrepreneurial cytogeneticists in the late 1960s sought a broader audience for their research. The XYY connection with crime seemed to offer such a chance, and at least initially, cytogeneticists welcomed the interest of the press. Only once public opinion became a hindrance to research did they find the media to be the culprit. These

issues came to a head in the newborn-screening programs and longitudinal studies that Court Brown and others had advocated and that were
being put in place from the late 1960s.

Longitudinal Studies and the Ethics of Research

Newborn-screening programs, under the lead of pathologist Neil Mac-
Lean, had been going on at Western General Hospital and other Edinburgh hospitals for some time. By the mid-1960s, MacLean and his colleagues, using the buccal smear technique for X-chromatin testing, had
screened thousands of newborns. When the program was in full swing,
about twelve hundred newborns in ten maternity units were being examined every month.[80] The studies were aimed to establish the frequency of
sex-chromosome abnormalities in the general population. In cases when
an unusual Barr body test was found, a full chromosome analysis was
performed. A full case history was collected and the data stored in the
Registry of Abnormal Karyotypes in the Edinburgh MRC unit for future epidemiological studies. A set of over 250 case reports of (anonymized) patients with sex-chromosome anomalies was also published in a
special report series of the MRC, to allow other researchers to draw on
the data.[81]

However, to identify the newborns with an XYY karyotype, a new
approach was needed. Building on existing links with hospital staff
and the uniquely equipped laboratory, the cytogenetics research unit
at Edinburgh started a new screening program based on full chromosome analysis of all newborns tested. All babies found to carry an XYY,
XXY, or XXX chromosome complement, together with a control group,
were enrolled in a longitudinal study to follow their development. Shirley Ratcliffe, an MD with a pediatrics specialization, was hired to direct the study. She was introduced to the new findings in human chromosome research early in her career when, in the 1960s, as a junior doctor
at the Hospital for Sick Children in London, she participated in a research project on cytogenetics and congenital heart disease. The longitudinal study in Edinburgh that she took up in 1968 and pursued until her
retirement in 1994 became her life's work.

Between 1967 and 1979, when screening stopped, 34,380 babies were
screened as part of the project. In the "blood laboratory" at the unit,
half a dozen trained technicians, mainly women, were devoted to karyo-

typing full-time. Among the babies screened, eighteen boys with an XYY chromosome complement were found and entered in the longitudinal study. Babies with an XXY and XXX chromosome set or other sex-chromosome anomalies were also enrolled in the study. The controls were recruited over a four-year period by asking the mother of the first child born each Monday morning to join the study with her infant. In total 271 children, including controls, were enrolled into the study.[82] Initially, parents were not informed about the reasons for the study other than its being a longitudinal study of growth and development. In the case of the XYY boys, however, the children's family practitioners were informed about the chromosomal findings. This arrangement was agreed to with the MRC, which funded the research. When the children were about four years old, parents were informed about the karyotype analysis, as it was thought that by that age they would have developed strong positive bonds with the children and could deal better with the information. From the beginning, the plan was to follow the children until adulthood. The babies and, later, the children and young adults were invited back to the clinic for checkups every six months. Ratcliffe and her team—consisting of an assistant pediatrician, a psychologist, and a nurse—recorded the growth of the children, their physical and sexual development, their speech development and cognitive ability (using the Wechsler Intelligence Scale for Children), and their psychological and social behavior as well as relevant life events. "Follow up was at a staggeringly high level," Ratcliffe remembered in a later interview:

> We tried very hard to feed back. The children came to the clinic, we fixed the time when they would come. Not like a hospital outpatient where always about one third of the patients didn't turn up. But we had the secretary to organize the attendance at the clinic. We provided transport if they needed it. We said no X-rays or blood tests unless they would be required in the ordinary way. And the child came in, we spoke to the mother and the child together. And then the child went and got measured and then they would see the psychologist and do the routine tests. And then we would sit all together and talk—anything. If the parents had any worries. And sometimes the worries were much stronger in the controls than in the cases.[83]

Ratcliffe deliberately avoided "publicity or premature publication of unwarranted assumptions," yet results of the study were regularly published in the scientific press.[84] Similar longitudinal studies started in other centers, notably in Denver under Arthur Robinson, and in New Haven, Toronto, Boston, Aarhus, and Winnipeg.[85] Some of these started a few years before the Edinburgh program, but none lasted as long or involved as many newborns. However, controversy quickly ensued. The Boston study in particular was singled out for critique, and this had lasting impact on the other projects as well. The arguments and counterarguments raised provide insights into emerging ethical discussions on clinical research and resistance to hereditary approaches to human behavior.

Stanley Walzer, a Harvard child psychiatrist, together with Park Gerald, a geneticist at the Boston Hospital for Women, one of the Harvard teaching hospitals, started a chromosome-screening project in 1968. The project was partially funded by grants from the National Institute of Mental Health (NIMH) Center for Studies of Crime and Delinquency. Unlike the Edinburgh study, the Harvard study screened only boys, parents were informed about the chromosomal findings from the very beginning, and no controls were included. Possible therapeutic interventions like psychological counseling and hormone treatment were also part of the program.[86]

From the early 1970s, apparently following the lead of child psychiatrist Herbert Schreier at Children's Hospital Boston, the study was strongly criticized by Science for the People, a group of science activists that had formed around protesting the use of science and technology in the Vietnam War.[87] The group was particularly strong in the Boston area, where, besides the XYY longitudinal study, it took a stand against plans to construct a high-security laboratory for work with recombinant DNA at Harvard's Molecular Biology Department, which led to a showdown at Cambridge City Council and a subsequent (short-lived) moratorium on research in the area.[88] Members of the group were also central to the defense of an environmentalist position in the race and IQ debate, which had been rekindled by the publication in 1969 of Arthur Jensen's controversial article on the topic.[89] Race issues were less central to the XYY debate, but race was the elephant in the room in the XYY-screening programs of institutionalized people, especially in the United States, where a large part of the prison population was black. A chromosome survey in institutions for "juvenile delinquents" in Maryland started in 1969 by cytogeneticist Digamber Borgaonkar, of Johns Hopkins Hospital, with a grant from the NIMH Center for Studies of Crime and Delinquency, was brought to a temporary halt by a press campaign and a lawsuit filed by the American Civil Liberties Union that exposed the unfair stigmatization of African Americans and the lack of informed consent.[90] More generally, the XYY debate raised similar issues as the perhaps more publicized IQ debate.

In 1974, Harvard microbiologist Jon Beckwith and MIT biologist Jonathan King, both members of Science for the People, published an article in the *New Scientist*, a magazine for general audiences, that exposed the "XYY syndrome" as a "dangerous myth" and singled out the Boston longitudinal study for critique. The article pointed to the problem of stigmatizing people and the dangerous use of doubtful "genetic information," for instance to provide evidence for continued incarceration or unsuccessful "therapy" attempts. Here the authors were most likely referring to the problematic attempts of Johns Hopkins psychologist John Money to treat XYY boys with anti-androgen drugs.[91] They criticized so-called research on the XYY syndrome as being "replete with biased, uncontrolled studies and extensive publicity for unfounded statements." Regarding the Boston study more specifically, the authors criticized the consent procedure, the "subtle coercion" of parents of XYY boys to enter the study, and the danger of "self-fulfilling prophesy" produced by telling parents of the XYY karyotype and its connected risks. This meant that the study was not just flawed and "worthless" but also "positively harmful" for the children involved. It also provided "the opening wedge for programs with much more serious eugenic implications." Beckwith and King cited the former president of the American Association of the Advancement of Science, Bentley Glass, who had expressed the hope that one day a combination of amniocentesis and abortion will "rid us of . . . sex deviants such as the XYY type."[92]

At the request of Science for the People, the Committee on Medical Research of Harvard Medical School convened a hearing on the XYY study. The question was also taken up by the then recently established Committee on Human Studies and the medical school faculty, and an informal opinion was solicited from the chief judge of the US Court of Appeals for the District of Columbia. Despite considerable debate, all three Harvard bodies voted in favor of the study's continuation.[93] The NIH also reviewed the study and renewed its funding for further three years.[94] Yet strained by the one-and-a-half-year battle, Walzer and his colleagues in 1975 ended the screening program (though not the longitudinal study). As reason for his decision, Walzer cited "the emotionally exhausting atmosphere" that included threatening phone calls and personal harassment, an accusation flatly denied by Science for the People.[95]

The debate around the Boston XYY studies coincided with federal efforts in the United States to strengthen regulations on human protection in research settings in the wake of the exposure of the infamous Tuskegee syphilis study—a forty-year study conducted by the US Public Health Service in which, even after penicillin was validated as a cure, treatment was withheld from nearly four hundred black men infected with syphilis to study the natural course of the disease.[96] The issue of interest in the XYY studies was that consent was at the heart of the debate, but at the same time meaningful consent was seen as rendering the studies useless. That the issue of patients' rights was played up against any form of experimentation and that "experimentation" was turning into a "dirty word" made even some patients' rights advocates wonder whether the pendulum had swung too far.[97]

The debate had raised intricate ethical questions on consent procedures, the right of parents to consent for their children, and the obligations to disclose chromosome findings. Yet Beckwith's main bone of contention was the attempt "to distinguish between the behavior of groups of people on the basis of genetics" and to focus "the blame for supposed antisocial behavior on the genes of the individual rather than on social, economic and familial conditions."[98] The stigmatizing and sensationalizing of the XYY genotype may have made this a special case and may have jeopardized the possibility of an unbiased study, but the "environmentalist" critique applied more generally.[99]

Critique in Britain was more subdued and late. However, by 1979, twelve years into the study, a series of damaging articles appeared in Scottish newspapers. "The Secret Guinea Pigs: Parents Not Told of XYY Tests" accused the *Evening Times*, "Scotland's greatest evening paper."[100] "Secret Tests on 14 Children with 'Criminal Chromosomes,'" the *Glasgow Herald* joined in.[101] Apparently "the story broke" when three parents were accidentally shown some doctors' notes revealing that their children were involved in the XYY survey. By this time the thirteen-part television series *The XYY Man*, adapted from a thriller series featuring the same main character by English writer Kenneth Royce, had turned the XYY karyotype into a household name, alarming parents about its possible implications.[102]

Ratcliffe and her colleagues tried to calm the waters, explaining the

legally backed decision not to tell the parents of the chromosome findings of their children, to protect the children and to not invalidate the
study. They also tried to dispel the false idea of a causal link between the
XYY karyotype, violence, and criminality, drawing on the preliminary
results of their own study and those of others. Reportedly, after talking
to Ratcliffe and her colleagues, the three parents who found out about
the purpose of the study agreed it would have been better for the project
if they had not known, and they consented for the project to continue.
Nevertheless, the issue was critically discussed in the Scottish commons,
and demands for new consent laws were raised. Under these circumstances, Ratcliffe and her colleagues felt compelled to end the screening.
The longitudinal studies at Edinburgh and a handful of other places continued well into the late 1990s, even if under more difficult conditions.
A series of workshops under the auspices of the March of Dimes Birth
Defects Foundation offered the most important forum for the groups to
meet and discuss methods and findings.[103] The studies were kept going
by the people who had initiated them and still believed in their importance, but they generally ended with their retirement or death.

Geneticists and physicians involved in the studies deeply deplored the
negative campaigns that brought the XYY-screening studies and most
longitudinal studies to a halt, especially in the United States, where
only the studies in Denver under Arthur Robinson continued. Besides
the specific tactics, including demonstrations and personal harassment,
that brought the Boston studies to a stop, they denounced the "geneticophobia" of the critics.[104] Jacobs elaborated: "It was inconceivable to me
and my colleagues . . . that there was a large body of professional people who did not and could not accept that both genes and environment
played an integral role in every aspect of biology, including human behavior. . . . None of the criticism came from human geneticists who
seemed genuinely shocked by the outrage being perpetuated on some of
their colleagues."[105]

The researchers involved in the investigations regarded the ending of
the XYY studies as a "tragedy," a disservice to those carrying the genotype, and an opportunity lost to achieve a better understanding of the relationships between genes and environment.[106] Indeed, the longitudinal
studies were producing results that helped counterbalance some of the
stereotypes of the XYY karyotype. For instance, Ratcliffe's initial studies pointed to some behavioral problems in some of the young boys with
an XYY karyotype. Yet with the first XYY boys entering puberty, the

studies highlighted the rather minor differences with the control group. Ratcliffe recalled: "Sometimes the worries were much stronger in the controls than in the cases. And, in actual fact, as you see from the last paper, they [the parents] weren't talking about criminality. The worst crimes were carried out by the controls. Not by the XYYs. . . . And this gradually became apparent."[107] Ratcliffe continued: "The findings were, in fact, very important for prenatal diagnosis. Because, originally . . . the attitude was—XYY, eliminate it. We cannot have these criminals around the place when we know they're going to be criminals. And that wasn't the case at all. . . . And in that respect, gradually, the obstetricians changed their attitude and said, you know, he'll be taller than the average boy. And he may have a little difficulty in school. But it's not to be severe and it can be handled in the same way as it is for a boy with ordinary chromosomes."[108] The question "What is to be done with the XYY fetus?" loomed large and was evidently of eminent importance at a time when more women made use of amniocentesis for prenatal diagnosis and new laws made abortion a legal option in Britain and elsewhere.[109] Ratcliffe took on an increasingly active role in combating what she saw as prejudice against individuals with an XYY karyotype. By the late 1970s, she made it clear that the rather slight differences in IQ scores found in her study did not provide grounds for selective abortion.[110] An educational leaflet for parents and families on the XYY condition, based on her work, was issued by Guy's Hospital Clinical Genetics Department and later updated by the Genetics Interest Group Scotland. It stressed that "men with XYY get married and have children just like men with XY chromosomes. . . . The majority of men with XYY live normal fulfilling lives and are completely unaware that they have an unusual chromosome pattern."[111]

A persistent problem remained that although the longitudinal studies did not show a significant association of the XYY karyotype with antisocial behavior in comparison with the controls, the studies in mental penal institutions did. The problem was discussed by Ernest Hook from the Birth Defect Institute in the New York State Department of Health who regularly compiled all available figures relating to the XYY karyotype.[112] A possible explanation was that the follow-up studies and early behavioral interventions in fact helped the children enrolled in the studies to not develop more serious behavioral problems. Another explanation was the small number of children enrolled in longitudinal studies, and consequently their low statistical force.[113] The discrepancy in the data nonetheless led to a division among practitioners according to their attitude toward the XYY fetus, with those looking at the adults with behavioral problems or learning difficulties having a more critical attitude than those following the longitudinal studies.[114]

Yet the XYY karyotype was not only linked with negative attributes. Researchers surmised that tallness and increased aggression connected to the extra Y could perhaps also have a positive impact on the competitive performance of XYY men. To investigate the hypothesis, researchers at Ohio University tested thirty-six basketball players for the presence of a double Y chromosome. None was found.[115] Unperturbed, researchers pursued the plan to screen athletes at the 1968 Olympics in Mexico City for the XYY karyotype. Apart from the interest in testing for the presence of individuals with an XYY karyotype among the highly selected group of athletes, the question was also raised of whether such a karyotype could provide an unfair competitive advantage.

The testing of the hypothesis became part of a broader genetic and anthropological study of the athletes launched by a group of local geneticists headed by Alfonso de Garay, the founder of the Radiobiology and Genetics Department at the National Institute of Nuclear Energy in Mexico City.[116] The project met with wide interest, and 1,265 athletes (30 percent of all those present) from 92 of 115 countries represented at the games volunteered to participate. The organizers hailed it "a study in human diversity."[117] Support came from the Organizing Committee of the Olympic Games, which included the genetic and anthropological study in the cultural program of the games, and from the National Commission of Nuclear Energy, which provided the Genetic and Human Biology Laboratory at the Olympic Village. Ford—from the Radiobiological Research Unit in Harwell—was in charge of the XYY-testing project and participated in two of the three preparatory meetings and at the games themselves, as reported in the British press.[118] Next to chromosome studies, the genetic program included the analysis of blood groups, blood proteins, PTC taste sensitivity, and fingerprints and palm prints.

No unusual sex-chromosome numbers were found among the male athletes. The Y chromosomes showed extreme variation in length, yet the differences could not be correlated to any sports specialty. It appears that the study ended there. No results were ever published, and the negative image of the men with an XYY karyotype prevailed, fueled at exactly the same time by the media interest in the Hugon and Speck trials and the murderers' presumed XYY karyotype. Although public attention around the XYY karyotype receded starting in the late 1970s, the negative image, once established, proved difficult to stamp out.

The XYY karyotype undoubtedly represented a particularly contentious case. Yet sex chromosomes were involved in other controversial practices. The same Olympic Games at which men were tested for the XYY karyotype also saw the introduction of compulsory gender verification for all women athletes based on chromatin testing followed by full chromosome analysis in unclear cases. The practice once more exposed the hopes placed in new genetic technologies and the challenges of relying on genetic explanations. This time, cytogeneticists were among the strongest critics.

Chromosomes at the Olympics

Compulsory gender verification in female athletes was introduced by the International Association of Athletics Federations at the European Championships in Budapest in 1966. The decision was driven by Cold War tensions between East and West that increasingly played out in the competitive sports arena. The tensions produced growing suspicions in the West that Eastern European female athletes were outperforming their Western counterparts because they were, in fact, men.[119] Drug controls were introduced at the same time to stop all presumed forms of cheating.[120] The gender test introduced in 1966 required female athletes to parade naked in front of a panel of physicians. From this practice it seemed a step forward when chromatin testing was first introduced on a small scale at the 1967 European track-and-field championship in Kiev. The International Olympic Committee followed suit in 1968.[121] At the Winter Games at Grenoble in France, only every fifth women was tested, but at Mexico City, all female athletes present, 781 overall, were tested. While this practice perhaps dispelled mistrust about the gender of certain athletes, it also led to deeply distressing experiences, as in the case of the Polish Olympic champion Ewa Klobukowska, who was the first athlete to fail the chromatin test in Kiev when she was found to have "one chromosome too many" to qualify as a woman.[122] Later reports suggested that she was an XX/XXY mosaic.[123] The occurrence of androgen insensitivity syndrome—women with an XY karyotype who are insensitive to testosterone levels in the body and thus do not have a competitive advantage over other women—further compounded the problems of relying on the chromatin test and karyotyping for gender identification.[124] Rather than representing a simple and straightforward test, it opened up new questions about who was to count as a woman, which criterion to apply, and what represented an unfair advantage in sport. At the same time, and somewhat paradoxically, the effort to define the possible competitive advantage that any of the known sex-chromosome anomalies might provide to female competitors reaffirmed the division of the sexes that governed the organization of competitive sports events.

From the very beginning, cytogeneticists criticized the use of chromatin and chromosome testing for gender verification and helped mount an increasingly determined campaign to abolish it. Among the earliest critics was Moore, who had worked with Barr on the development of the chromatin test. Addressing the issue in an editorial of the *Journal of the American Medical Association* before the Olympics, he pointed out that "buccal smears, reflecting chromatin or nuclear sex, or chromosomal analyses, indicating chromosomal sex, cannot be used as indicators of 'true sex'" and thus should not be relied on for gender verification at the Olympics.[125] As he noted, the test had long been abandoned in the clinical context for gender determination. Another early critic was Malcolm Ferguson-Smith from Glasgow University. Invited by the British Olympic Committee to administer the tests for athletes participating at the Commonwealth Games in Edinburgh in 1970, he declined, arguing that the chromatin test was "not the appropriate test." On the contrary, it would be more likely to detect and unfairly exclude female athletes with androgen insensitivity than males masquerading as females. The responsible medical adviser countered that, though "not infallible," the test was generally regarded as "sufficiently accurate" for the purpose.[126] As it turned out, androgen insensitivity syndrome affected about one in every five hundred athletes, whereas thirty years of testing did not uncover a single genuine male impostor.[127]

Confronted with the decision of the International Olympic Committee to continue the practice of chromatin testing, Ferguson-Smith and others critical of the system later agreed to carry out the test, to ensure that athletes who failed it were promptly and adequately informed of its possible pitfalls as well as of their right to compete regardless of the test result, as in cases where an androgen insensitivity syndrome was confirmed. Other cytogeneticists, including, for instance, Jacobs, flatly refused to be involved in any way in gender verification.[128] Women athletes on the whole supported the practice as in their eyes it deterred cheating. Critics argued that the refusal of the International Olympic Committee to release any data about the testing program meant that most athletes were not aware of the serious problems produced by the testing regime.[129]

Ferguson-Smith, who became a member of the International Amateur Athletic Federation's Working Group on Gender Verification, founded in 1991, and other critics became convinced that all sex testing in international sport events should be abolished because it was demeaning to women, unreliable, and, in a few cases, seriously damaging, apart from being expensive. They also argued that the existing antidoping tests, especially the requirement to provide a urine sample under direct visual observation, made a gender test redundant.[130]

Despite this sustained critique, compulsory gender verification in the form of buccal smear testing followed by full chromosome analysis and measurement of blood hormone levels in negative or inconclusive cases, remained in place until 1991, when increasing pressure led to the discontinuation of the practice in the Olympics. It was reintroduced at the 1996 Olympics in Atlanta, using the polymerase chain reaction to test for Y-chromosome material rather than the chromatin test. It was again suspended shortly before the games in Sydney in 2000. The reasons for eliminating the screening rested on changing sensibilities as well as the increasing realization of the limitations of the genetic test. Nevertheless, chromosome tests still play a role in the battery of gender verification tests applied in specific cases. As recent examples show, humiliating and damaging situations persists for women whose gender is called into question, and individuals not falling into one or the other gender category according to present criteria are still banned from competing.[131]

A Genetic-Testing Culture

Following the X and Y chromosomes through mental hospitals and prisons, the courts, maternity wards, genetic counseling clinics, and the Olympics, this chapter has traced how specific technical approaches and their widespread use, multiple interests, and debates propelled and kept sex chromosomes at the center of attention. Researchers studying sex chromosomes and people who were picking up the techniques pursued questions that reached far beyond issues of gender identification and gender differences. They examined the genetic basis of mental disability, the role of genetic predispositions to socially deviant and criminal behavior, and the role of sex chromosomes in competitive sports performance. They moved from fundamental questions about the structure and functions of genes and the interaction of genes and environment to eminently practical questions of clinical diagnosis that could guide interventions and treatment and shape patients' identity. Yet rather than providing straightforward answers or solutions to long-standing questions, practices around human sex chromosomes remained entangled with existing attitudes about sex and gender and unfailingly opened up more questions than they set out to answer. They met with wide scientific and political resistance to privileging biological over social determinants of human behavior and increasingly came under ethical scrutiny.

The long controversy around the XYY karyotype, spanning from the mid-1960s into the 1990s, when the last longitudinal studies ended, provided a particularly useful window for capturing the multiple practices, changing attitudes, and social and ethical concerns in which research and practices around sex chromosomes and human heredity became enmeshed, and so kept alive the controversy around the XYY karyotype. The original observations that focused attention on the XYY karyotype resulted from the expanding screening programs under way from the late 1950s that traditionally targeted institutionalized populations like prisoners and patients in mental institutions as well as, increasingly, newborns. The findings around the XYY karyotype, in turn, stimulated new large-scale screening programs involving tens of thousands of individuals. The presumed association of the XYY karyotype with aggression and crime resonated with ongoing debates and anxieties over increasing violence in the public sphere and, in the context of spectacular murder trials, catapulted sex chromosomes onto the front pages of various media outlets. The subsequent neonatal screening and prospective studies set in place to clarify questions raised by the early findings became a testing ground for the incipient bioethical debates of the early 1970s. New requirements for informed consent challenged the design of screening and follow-up studies, while genetic testing, especially in the new context of prenatal diagnosis, led to troubling questions about which information should or could be communicated to would-be patients or pregnant women, given that its meaning was unclear and a wide spectrum of outcomes was possible. These and other issues were taken up by new advocacy groups such as Science for the People that more generally challenged the search for biological determinants of social ills.

Together these aspects go some way toward explaining the length and vehemence of the debate around the XYY karyotype.

Yet despite the complications of and resistance to sex-chromosome research, the continuing focus on these issues meant that more people became accustomed to genetic testing, chromosome images, and the possible meanings of the results, setting the stage for an increasingly widespread and accepted genetic-testing culture.[132] The prevailing negative attitude against genetic explanations of human behavior also shifted considerably in the following decades, not least because of the expanding use of karyotyping technologies. The debate around "criminal chromosomes" is now mostly relegated to a bygone dark era of genetics, yet the search for biological determinants of criminal behavior, situated in genes or the brain, is an ongoing concern.[133] Similarly, many of the questions posed in the course of the debate around the XYY karyotype are being raised again today in the context of new technologies for genome-wide DNA sequencing of individuals.[134]

The following chapter more closely considers the population-based chromosome-screening studies undertaken from the 1960s that underpinned some of the sex-chromosome studies but also tackled other questions, such as the study of genetic variation in human populations. These efforts were supported by international organizations such as the WHO and marked yet another field in which chromosome researchers hoped to make an impact. The fifth and last chapter ponders in more detail the long-term legacies of the human chromosome studies of the 1960s and 1970s and their relation to the molecular studies of human heredity in light of more recent developments.

CHAPTER FOUR

Scaling Up

The fascination with chromosome pictures very much relied on the fact that each preparation represented the genetic makeup of one individual. Again and again, scientists, physicians, patients, artists, and "sitters" recognized the pictures as individual "portraits." They were pasted into clinical records, functioned as personal business cards, and, in artistic variations, went on display in portrait galleries.[1] Yet the same techniques of karyotyping were also employed in large-scale population studies of newborns, prisoners, and patients in mental institutions and, in principle, if not in practice, served to surveil a world population exposed to nuclear radiation. This prompts the following questions: What facilitated the scale-up of the technology? How were populations defined? Why did they become the focus of sustained investigation? This chapter addresses these questions by introducing populations as a central topic for a broad range of scientific and political concerns in the postwar era. It then analyzes a series of tools and infrastructures, including blood preparation techniques, registries, and efforts to develop pattern-recognition software that facilitated the scaling up of human chromosome research. Finally, it investigates in more detail some of the chromosome studies of clinically and geographically or culturally defined populations and their continuing legacies.

The efforts to automate chromosome analysis point to the difficulty of replacing the human observer and thus once more highlight the visual skills needed to sustain the microscope-based study of chromosomes. Intriguingly, the most detailed descriptions of the steps involved in analyzing chromosomes under the microscope can be found in the literature on automation. At the same time, not all population studies undertaken with the karyotyping techniques involved big numbers, although they always referred back to larger data collections for comparison. In some cases, the scaling up could be geographical and bridge temporal scales, as in the effort to test people in the most remote and isolated communities in the world, who were expected to hold the key to the deep evolutionary history of humans. International organizations, including the WHO and the International Biological Program (IBP), organized by the International Council of Scientific Unions, provided the necessary infrastructural support for these projects.

Focusing on the population studies undertaken with the new karyotyping techniques, the chapter explores a further broad set of technical, methodological, ethical, and political questions and contentions that shaped the field of human heredity in the long 1960s. It emphasizes the continuing role of concerns and opportunities around nuclear radiation, the buildup of large data sets and other infrastructures, the relentless search for genetic differences in human populations, and the role of international organizations in sustaining research agendas and carving out a space for human heredity in the postwar era.

Moving Populations Center Stage

Since the 1920s human populations had been a focus of study and intervention for a wide range of fields, including demography, anthropology, epidemiology, and public health.[2] Heredity played a role in many of these efforts. Anthropologists were seeking to establish heritable biological differences between human populations, while eugenic thinking made heredity a central concern for researchers, administrators, and politicians engaged with epidemiology, public health, demography, and economics. At the same time, population geneticists were developing the statistical tools necessary to study the distribution and transmission of heritable traits in animal and human populations. After 1945 these efforts continued but under changed scientific and political constellations. In particular, the focus on human populations was heralded as a way to disengage from the racial thinking of the prewar era, even if recent scholarship points to considerable overlaps between typological and population-based thinking, both before and after the war, in Germany and more generally.[3]

To study human populations, chromosome researchers combined their karyotyping techniques with epidemiological approaches. Their aim was to establish the "normal" frequency of chromosome variation in the general population as well as the association of chromosomes with disease patterns or environmental risks. A population thus could be defined epidemiologically, including control cases to avoid ascertainment bias, or it could be taken to be representative of the "population at large."[4] Chromosome studies also piggybacked on anthropological investigations that studied human variation and established evolutionary patterns in human populations. Yet before looking into some of these studies in more detail, it is necessary to consider the tools that made it possible to scale up the study of human chromosomes.

Collecting Data

In the late 1950s and early 1960s only a handful of centers around the world had the necessary skills and equipment for human karyotyping. Among these, the institution that most vigorously pursued the use of karyotyping techniques for large-scale epidemiological studies was the MRC Clinical and Population Cytogenetics Unit in Edinburgh, headed by Court Brown. Embracing the study of human populations, Court Brown combined new cytogenetic techniques with the epidemiological approaches that had characterized his earlier work on the relation of radiation treatment and leukemia, as well as other studies he pursued in collaboration with the epidemiologist Richard Doll.

As already mentioned, human chromosomes became amenable to large-scale epidemiological studies only after Nowell and Hungerford had shown that white blood cells extracted from a blood sample could be used for karyotyping. Before then, chromosomes were mostly prepared from testicular or bone marrow samples, both of which had to be operatively extracted. These were hardly procedures that could be suggested to healthy individuals for unbiased epidemiological studies. But even before a less intrusive skin culture method and then the peripheral blood method came into use, Court Brown, setting up a human karyotyping laboratory at his unit at Edinburgh, was thinking in epidemiological terms.

Already in 1959—directly following the first reported chromosome-linked developmental disorders and well before the announcement of the new blood culturing techniques for chromosome studies—he started the Registry of Abnormal Karyotypes to gather data on a large scale and open it up for epidemiological studies. We have already considered the registry as a tool to engage clinicians in the new project of studying chromosome diseases.[5] Here I would like to highlight the immense work that went into gathering the data and maintaining the database that became an essential tool for the epidemiological work in the unit and beyond. Data collection and record management, notably in asylums and later in specialized institutions such as the Eugenic Record Office at Cold Spring Harbor, lay at the heart of much early work on human heredity and continue to play a role in the genomic era.[6] These tools have their history and their own materiality. They are revealing for the kind of data they combine, as well as for the practices that sustain them and the questions they generate. Furthermore, the controversies that occasionally flare up around the apparently rarefied data open a window on evolving ethical discussions regarding work with human subjects.[7]

Data collection for the Registry of Abnormal Karyotypes in Edinburgh was extensive. Each entry in the registry consisted of a large index card and linked files that combined karyotype and comprehensive clinical data, supplemented by family data. The clinical information was compiled from a four-page form that asked for personal data; medical history; a detailed description of physical characteristics, such as body measurements, weight, hair, voice, and sexual traits; physiological data ranging from blood counts to color vision; intelligence tests results; family data; and reproductive history.[8]

Various people contributed to data collection. Most cases were identified through surveys performed by researchers in the unit. Collaborating physicians supplied additional cases. Clinical specialists provided extensive case descriptions as well as tissue samples. Researchers and technical personnel performed the chromosome analyses. Finally, scientifically trained assistants initiated contact with the families and collected family histories of diseases. They constructed pedigrees and checked information on birth, death, and marriages against the records of the General Register Office, the Central Office for National Health Service Records, and census data. Their work also included annual follow-ups with individuals in the registry through their attending physicians and record management. The unit hired a statistically trained epidemiologist to advise on how to implement the data to facilitate statistical analysis. Researchers at the Galton Laboratory in London helped with the pedigrees and linkage studies.[9] Data collection focused on Scotland, with the expectation that it could be expanded to cover the national territory. Yet there was an advantage to work in Scotland, as public records there included information that greatly facilitated genetic research. For instance, unlike in England, the birth certificate listed the date and place of the parents' marriage. This information could be used to trace the parents and grandparents and their records.[10] At least initially, the registry was accompanied by a sample collection, although reference to it is scarce.

By 1962 the registry contained over two hundred cases. Court Brown reckoned that it was "not an exaggeration to say that the data already in the registry was unique in terms of their amount, their quality and, for sex chromosome abnormalities, their scope." He was convinced that the information "could well be of the greatest value to geneticists, other cytogeneticists, endocrinologists and human biologists."[11] The entries do not seem to have been anonymized. Nevertheless, the registry was open to everyone in the unit and other bona fide researchers. Court Brown, always eager to advance the cause of cytogenetics, was keen to share the collected data even more widely and convinced the MRC of his plan. A volume published in the MRC Special Report Series in 1964 introduced its readers to karyotyping, the newly identified genetic diseases, and the registry and presented full case reports of 266 anonymized patients with sex-chromosome abnormalities extracted from the registry. In the preface, MRC officials expressed the hope that the volume would "serve not only as a useful handbook for workers in many different fields, but also as a source book of data for study and analysis and as a stimulus to speculation and further inquiry."[12]

By 1965 the registry contained the data of more than eight hundred individuals. It continued to grow at a rate of 100 to 150 cases a year and soon occupied two small rooms in the unit. At the peak of activity two full-time and two half-time assistants worked on the records. Anna Frackiewicz was in charge until 1968, then Susan Collyer, and later Rhona De Mey took over. They developed great skill in using public records for genetic research and efficiently ran the registry.[13]

Through inclusion in the registry, single patients, whose chromosome variant might have been discovered through individual diagnosis in a clinical context, were enrolled in population-based epidemiological studies. Court Brown envisioned that cytogenetic facilities would be made available within the framework of the recently established National Health Service in Britain and that the data acquired would be centrally pooled to study the public health dimension of the new set of genetic diseases. Indeed, he anticipated that Edinburgh could become the data-processing center for this endeavor.[14]

By the early 1970s, several cytogenetic registries existed in various countries. At a meeting in Edinburgh, organized by the Standing Committee of the Paris Conference on the Standardization in Human Cytogenetics, the possibility was discussed of merging these into one central registry. In the end the committee decided that the time was not quite ripe, but it highlighted the need to standardize the way information was collected and recorded. The committee also solicited cooperation between the different institutions and the international support of the WHO in the endeavor. By this time, the registries were expected to serve vastly expanded aims, ranging from morbidity and mortality studies to recurrence risk and reproductive fitness studies, estimates of mutation rates, etiological studies, determination of karyotype-phenotype correlations, linkage studies, the determination of breakpoints in structural rearrangements, and the determination of health-care needs of patients with chromosome anomalies.[15]

Data were added to the Edinburgh registry well into the 1990s, but by the end of the decade, the registry closed down. Exactly what caused the decision and the actual fate of the records remain unclear. Although people are reluctant to speak, it appears that it was not a specific incident but rather changing scientific priorities and especially new ethical guidelines that made the registry seem not only outdated but also outright problematic.

From the 1980s molecular biologists started being hired in increasing numbers into the unit. The data collected in the registry proved supple enough to offer useful starting points for molecular research. This was especially true for the genetic data on families and the extensive information on chromosomal translocations that could be exploited to identify and locate genes with the new molecular mapping technologies (see chapter 5). Nevertheless, it soon appeared that the most promising case histories had been skimmed off the top, and the registry increasingly appeared a relic from a bygone era.

More decisively, perhaps, in the wake of public uproar following the discovery of the unauthorized removal and retention of children's organs at two hospitals in Britain in the mid-1990s, the MRC tightened its ethical guidelines on the collection and use of human tissue and personal medical information. Above all, it affirmed the need for patients' consent and the principle of confidentiality. The new guidelines were circulated to all MRC-funded researchers.[16] The whole affair was widely reported in the press. This will undoubtedly have raised concerns regarding the information held in the registry. It may explain why people in the unit reportedly "wanted to bury" the registry and why, until today, they are reluctant to speak about it or acknowledge their connection with it in any way.[17]

What actually happened with the registry remains unclear. Most people seem to believe that the registry is still kept somewhere in long-term storage, although nobody seems to know where. One well-informed person knew positively that the registry had been digitized for archival reasons, a possibility denied by everyone else. Most tantalizing, one person produced a few index cards apparently from the original paper registry, filled out in a neat hand, but was nervous to show them for longer than a few seconds. The problem was a lack of consent, but even destroying the records would require approval from an ethics committee. For this reason, the registry, according to this informant, was in limbo. Reconsenting it did not seem worthwhile, although it still contained useful information that could not be gained from molecular technologies alone. The same person also volunteered the information that an anonymized Trisomy Registry started by Patricia Jacobs at a later date was active until 2004. Describing that registry as a "fantastic resource," the researcher confirmed that chromosome registries contained much information that could be of interest to molecular biologists. To show that "one never knows what might turn out useful," the researcher cited the example of the "beautifully documented" Scottish Mental Survey, an educational survey of 1947 that built on an earlier such survey of 1932. The surveys were "stuck in the archive" and forgotten until a researcher found them, contacted the individuals, and retested them, creating a unique long-term data set for the study of cognitive development. Against this reasoned version of events, another equally well-informed source in the same institute assured me that the index cards had been given to a commercial firm for shredding. Only the medical records were kept as the regulations demanded.[18] Of course, these stories are not necessarily mutually exclusive and a combination of various aspects might be true.

The fact that the records are not available limits our understanding of how the registry worked. How was the information organized? Who filled out the cards? Were they annotated? Which finding aids existed? How could statistical information be retrieved? Even if it is not possible to answer any of these questions, the fact that the registry remained in operation for over forty years points to its central role for the unit's work on human population cytogenetics.

Given its importance, it may be surprising that the registry, unlike, for instance, McKusick's catalog of *Mendelian Inheritance in Man*, was never digitized. This may have contributed to it becoming outdated as a research tool.[19] And yet a lot of effort in the same research establishment went into harnessing computers to automate the analysis of chromosome pictures.

Automating Karyotyping

Early on, Court Brown realized that the time needed to count and analyze chromosomes posed the most serious restriction on the development of population cytogenetics as he envisaged it. To cope with the ever-larger number of samples in the studies pursued in his unit and to alleviate the tediousness of karyotyping, he placed high hopes in an ambitious project to automate karyotyping. He presented the plan to the MRC thus: It takes more than one year to train a technician to the point that he or she—in most cases the technicians were women—can count and analyze one karyotype in about four minutes. A trained technician could analyze a maximum of fifty cells per day. A second observer had to check the results. Automation would speed up karyotyping to one minute per cell count and hence increase the overall number of cells counted from seventy-five thousand to five hundred thousand per year.[20] This would allow for a considerable scaling up of the screening projects. In addition to speeding up karyotyping, researchers hoped that computerized methods for measuring the optical density of the chromosomes might be applied to find aberrations that could not be observed with the light microscope. The method would quasi "weigh" chromosomes rather than measure their length. In short, there was hope that the machine could add accuracy as well as speed.

Discussion on the automation project in Edinburgh started in the early 1960s, in connection with Court Brown's work on a chromosome-based biological monitoring system for radiation exposure (chapter 1). The prospect of setting up such a system was as exciting as it was daunting. There was little doubt that monitoring for radiation as well as for workplace pollutants like benzene was "critically dependent" on automated techniques for counting and analyzing chromosomes. As Court Brown explained to the secretary of the MRC, with whom he was in close contact about the radiation work, "We want to be in a position rapidly to count and analyze 500 cells per person, and this is altogether too much for the human being [looking] down a microscope."[21] In theory, the problem could be solved by employing a large number of people, but this only exacerbated the problem of having to deal with personal differences in scoring cells. Moreover, Court Brown was categorically opposed to setting up a "sweat shop" for karyotyping.[22] The only way forward was automation, and Court Brown was already in conversation about possible solutions with the British electric engineering and computer firm Ferranti; the electronic division of the Atomic Energy Authority, which was keenly interested in the problem; and the University Computer Unit. Everyone seemed to agree that automating karyotyping should be possible, even if writing the computer program would be challenging. "Needless to say," Court Brown warned his interlocutor, "this is all going to be very expensive, and figures in excess of £100,000 are now quietly but casually being mentioned."[23] Other population studies pursued by researchers in Edinburgh added to the strain on resources and made automation seem not just desirable but imperative.

On the other side of the Atlantic, in Washington, DC, the physicist and computer enthusiast Robert Ledley at the National Biomedical Research Foundation, a nonprofit organization he had founded, had already made headway in that direction. Dedicated to the problem of developing automated devices for pattern recognition in the laboratory, Ledley had assembled a film input to digital automatic computer, known by the acronym FIDAC (fig. 4.1). The machine was designed to scan a photomicrograph at high speed and high resolution and to feed the information to a large computer that would recognize patterns. Ledley saw many possible applications for his film-reading device. Chromosome analysis was "one of the most important" ones—presumably because of its broad applications but perhaps also because, as another pioneer of computer-oriented image analysis put it, "the highly stereotyped, simple, stick-figured chromosome was ideally suited to the fledging capability of this young technology" (fig. 4.2).[24] Collaborating with the cell geneticist Frank Ruddle from Yale, Ledley developed the software FIDACSYS, which was able to analyze chromosome images fed into a computer by FIDAC. The software distinguished single chromosomes with a system based on boundary recognition and calculation of the centromere index

FIGURE 4.1. The FIDAC scanner for automatic pattern recognition developed by Robert Ledley, hooked to an IBM 7094 computer (foreground) at the Goddard Space Flight Center outside Washington.

Source: Ledley and Ruddle, "Chromosome Analysis," 43. Reproduced with permission of Scientific American.

FIGURE 4.2. Photomicrograph of chromosomes displayed on the FIDAC monitor.

Source: R. Ledley et al., "Progress Report on Biomedical Picture Data Processor," NIH Grant No. GM 10797, 1 August 1964. NBR Report No. 64081/10797, Silver Spring, MD: National Biomedical Research Foundation; Ledley Papers (uncataloged), box 43, file Biomedical Pictures Data Processing. NIH GM10797, 8/1/1964, NLM. Reproduced with kind permission of Fred Ledley.

(relation of length of the short arm to full length of the chromosome) (figs. 4.3 and 4.4).[25]

Ledley presented his apparatus, including the software, at the 1964 conference "Mathematics and Computer Science in Biology and Medicine" organized by the Medical Research Council in association with the

FIGURE 4.3. Francis H. Ruddle (center) and Herbert A. Lubs in conversation with a technician using FIDAC.

Source: "Spotting Flaws in Genes," 23. Photograph by Gene Daniels, further credited to A. Robert Street. Courtesy of the McGovern Historical Center, Texas Medical Center Library, IC077 Medical World News Photograph Collection, IC077-7063-001. Reproduced with permission of McGovern Historical Center, Texas Medical Center Library.

FIGURE 4.4. Printout of chromosomes as stored in computer memory.

Source: Bauer, "A Letter from the Publisher," 19. Courtesy of the McGovern Historical Center, Texas Medical Center Library, IC077 Medical World News Photograph Collection, IC077-7063-002. Reproduced with permission of McGovern Historical Center, Texas Medical Center Library.

British health departments in Oxford, where it produced some buzz.[26] A few months later, Patricia Jacobs from the Edinburgh group, together with a representative from the Atomic Energy Authority, traveled to Washington to inspect the scanner. They were sufficiently impressed with the machine to recommend acquisition of a FIDAC. They saw some drawbacks—that the machine was reading from film rather than directly from the microscope and that it needed to be hooked up to a powerful computer for operation. Few institutions could afford to have their own computer, and negotiating access to other computers for any length of time was difficult and costly.[27] Yet the British researchers hoped either that the machine could be adapted to their needs or that experience with FIDAC could help design and build a better machine. The MRC, which was open to investments in the field, as evidenced by its sponsorship of the conference that had brought Ledley to Britain, approved the proposal, and arrangements were made for a FIDAC to be shipped to the United Kingdom.[28] The machine was installed at Imperial College London, one of the few institutions in the country that housed an IBM 7090 computer. Denis Rutovitz, who was trained in pure mathematics but also had some background in computing, was hired to do the development work. He headed a small team known as the Pattern Recognition Group. Though located in London, the group formally belonged to the Edinburgh unit. Eventually it moved to Edinburgh, where people and hardware filled a whole floor of the new three-floor institute building.

Court Brown remained restless. It would be two or three years, he complained, before it would be clear whether FIDAC did the work they expected from it. Giving the importance and demand for chromosome work, was there not scope for a grander scheme, he asked, in which the automation of chromosome analysis would function as a "prestige operation" in the application of pattern recognition in biology more generally? Court Brown envisaged the creation of a dedicated computer cytology unit, staffed with computer scientists and statisticians who worked in close collaboration with biologists and clinicians.[29] At minimum he wanted to see a "crash programme" in place to perfect automation using FIDAC.[30] Yet to his chagrin, things did not move quite as fast as he would have liked.

In the 1960s there was considerable momentum with regard to the introduction of computers in biology and medicine, and specifically in the study of heredity. The conference in Oxford to which Ledley was invited was just one of a series of conferences and workshops dedicated to the topic, and various institutions, notably the NIH but also the WHO, actively promoted the introduction of computers to biomedical research and medical practice.[31] In human genetics computers were harnessed to collect and organize data on genetic diseases, to calculate genetic linkage, and to simulate gene flow through populations. Pattern recognition, too, was a busy field, with several groups making contributions. However, pattern recognition proved a difficult problem to tackle with a machine.

Experimenting with FIDAC provided Rutovitz and his team with "invaluable" experience.[32] Yet even before the researchers had tested FIDAC's effectiveness for population screening, they decided that using a film scanner was not practical, as working from film not only was "superfluous and time consuming" but also restricted the analysis. An obvious drawback, but by far not the only one, was that it meant losing the capacity to alter the focus of the microscopic image. This, in turn, led to a considerable loss of information.[33] To gain higher definition, it was necessary to work directly down the microscope. This was consistent with the established practice in the Edinburgh unit to analyze chromosomes under the microscope rather than working with photographs, a method that Jacobs described as "incredibly tedious and time consuming" with "little to recommend it other than the production of a large number of pictures suitable for mural decoration."[34] More generally, Rutovitz and his team found that many of the claims Ledley had made about his apparatus at the Oxford conference were "misleading" and that FIDAC was not able "to do anything like a complete job."[35] To move forward, Rutovitz sought support for the acquisition of a "Computer Eye," a sophisticated if costly computer scanning and display system produced by Information International Inc. of Los Angeles that could be used to work straight from the microscope image, as well as a small dedicated IBM computer. Although positively impressed by Rutovitz's work, the MRC became concerned about the risks involved in a "large scale venture in the automation field."[36] To make sure that "the magnitude of the financial outlay" was matched by the "possible scientific dividend," MRC officers looked for assurance that automation was a practical proposition and that the potential of chromosome analysis was not overrated.[37] The question became entangled with discussions on the long-term future of the Edinburgh group that looked into expanding and getting new accommodation. Having convinced itself that "the field [of chromosome analysis] would not 'dry up' in the foreseeable future" and that "even if the whole chromosome story were to fall down, there would still be a need for large-scale pattern recognition techniques" in other medical fields, the MRC gave Rutovitz the go-ahead.[38] Work on FIDAC continued, but much of the development work shifted to the Computer Eye system, which was installed in a space close to the MRC's own new Computer Centre in London until new space would become available in Edinburgh and the group could acquire its own computer.[39] Rutovitz expected that progress with the Computer Eye would soon overtake that made with FIDAC. Yet the path proved "long, very long."[40]

The idea was first to work at low resolution to find metaphases or dividing cells in the samples with the automated scanner and then to focus down and perform an automatic chromosome analysis (fig. 4.5). This procedure closely followed the working practice of the human observer. To accomplish the task, the group collaborated with an engineering firm in Royston (Metals Research Ltd.) to develop a mechanically driven scanning microscope that could be used in conjunction with the Computer Eye. Hooked up to the computer, the microscope would move along the microscopic slide, search for metaphases, and center and focus the preparations for analysis. Chromosome analysis was translated into the following steps: digitization of the visual image, field segmentation, feature extraction, and classification (fig. 4.6).[41] Instead of the usual pixel-by-pixel scanning of the image, the machine used an interval-coding scheme developed in Edinburgh, which slimmed down the data significantly. All the other steps also required innovative solutions.

Pattern recognition for chromosomes related to several other problems that computer scientists were tackling at the time. Prominent among these was the challenge of character recognition for both typescript and handwriting. This was of interest, for instance, for digital copying, and there were commercial interests involved. The connection between character and chromosome reading is intriguing, as chromosomes were often compared to letters, metaphorically and otherwise. Yet chromosome recognition turned out to pose even bigger problems than character recognition. Chromosomes could overlap and lie in different orientations, and their distribution seemed random. There was no measuring involved in character recognition and no interest in the density of letters or in translocations, as was crucial in chromosome analysis. Different algorithms were developed to overcome these problems,

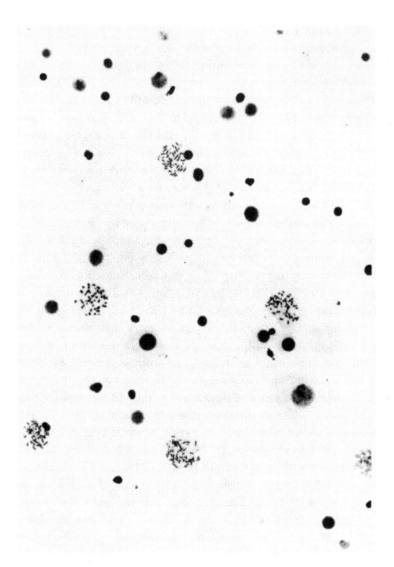

FIGURE 4.5. Microscope slide with chromosome preparation. Some cells are fixed during cell division when chromosomes are visible and amenable to chromosome analysis. Other cell nuclei are in the resting stage.

Source: Piper et al., "Automation," 204, fig. 2. Reproduced with permission of Elsevier and of Jim Piper.

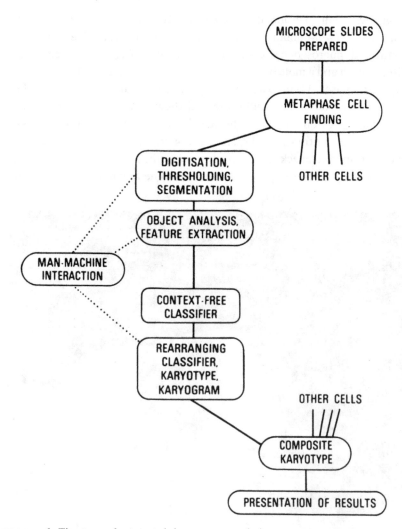

FIGURE 4.6. The stages of automated chromosome analysis.

Source: Piper et al., "Automation," 207, fig. 4. Reproduced with permission of Elsevier and of Jim Piper.

yet despite large investments in the field, humans continued to outper-form machines. As Rutovitz put it in a later interview, "Human eyes can do tricks that are damn difficult to mimic with machines."[42]

The solution to this problem was sought in an "interactive system" that combined the scanning and measuring capacity of the computer with the pattern-recognition ability of the human operator. The machine the group eventually developed and successfully commercialized did exactly that. It was a self-contained machine with a small interactive (computer-controlled) microscope and a screen-based editor (fig. 4.7). It used a scanning camera and a motorized microscope stage to scan the slide and identify metaphases. Using pattern-recognition software, the machine then sorted the chromosomes and presented them to the human operator in a paired display on the screen (fig. 4.8).[43] The human operator "assisted" the machine in "tricky" tasks, such as sorting out chromosome clusters and overlaps and checking the machine classification. An innovative system allowed the user to point to a chromosome and move it away. Cyto-

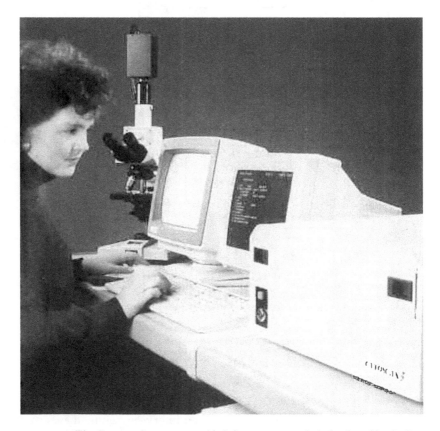

FIGURE 4.7. The Cytoscan for computer-aided chromosome analysis developed by the Pattern Recognition Group in Edinburgh. The machine included a computer-linked microscope and a computer screen for editing.

Source: Piper, "Cytoscan," 27, fig. 3. Reproduced with permission of the Medical Research Council and of Jim Piper.

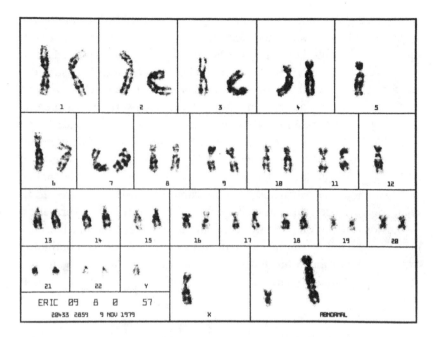

FIGURE 4.8. A karyotype of banded chromosomes automatically produced from a digitized cell image. Two abnormal chromosomes resulting from a translocation between chromosomes 5 and 12 are presented in the bottom-right corner of the display.

Source: Piper et al., "Automation," 205, fig. 3. Reproduced with permission of Elsevier and of Jim Piper.

scan, commercialized by Image Recognition Systems, a UK firm, became the most successful machine of its kind. By the early 1990s, more than one hundred laboratories in twenty countries used the machine.[44]

The original problem that Court Brown was hoping to solve—the automation of karyotyping for radiation and environmental monitoring—needed to wait a little longer to see a solution. In this type of work, the incidence of mutagenic events was so low and the number of metaphases to be analyzed so high that it was not feasible for a human operator to interact continuously with the machine. For this task the Edinburgh group developed a system that detected the rare events and presented them collectively to the operator for evaluation. Even if the system was not completely automated, it led to a "massive reduction" of "human time" that needed to be devoted to the analysis.[45]

Despite this eventual success, Rutovitz came to quite a sober assessment of his life's work. He conceded that, "had Court Brown lived, we would have disappointed him," although "perhaps with his direction the work would have developed differently."[46] The problem was that development never quite kept up with the scientists' computing needs, and researchers found other ways of doing what they wanted. One such development was the introduction of chromosome-banding techniques that made many existing automated systems "obsolete overnight" while giving cytogeneticists new tools that much diminished their dependency on machines.[47] There was also some ambivalence in the way things were set up in the Pattern Recognition Group in Edinburgh, as "the masters were research people but the clinical applications offered a bigger economic return." At a time when the MRC became increasingly interested in technology transfer, this was an important consideration.[48] Nevertheless, throughout the period there was constant development and much excitement in the field of automated pattern recognition and the continuing expectation that an automated system would become available allowed the Edinburgh unit to entertain an ever-expanding program of population studies.[49]

Normal and Clinical Populations

The first and in many ways most fundamental population study the Edinburgh unit embarked on was the newborn-screening program. The unit started screening newborns in 1959, very shortly after the announcement of the first chromosome anomalies. The project was based on the conscious decision to link the study of chromosomes to epidemiological techniques.[50] While initially screening was based on Barr body testing, from the mid-1960s a new testing program started including full chromosome analysis for all babies who were tested. This later project went on for over a decade and involved tens of thousands of newborns. At the height of the project, all babies born in the two major maternity wards in the city, situated in Western General Hospital and Eastern General Hospital, were tested.

The screening of newborns, one of the first of its kind,[51] was meant to establish a baseline for the incidence of chromosome mutations in the general population. This was of crucial importance for the calibration of other studies performed on selected populations. In addition, the newborn-screening program was aimed at identifying infants that might need special attention and at providing data for genetic counseling. The long-term goal was to correlate the chromosome anomalies with parental age, social class, and ethnicity and thus to identify the causes of the anomalies.[52]

A first report on the Edinburgh study published in 1961 presented the results from six thousand newborn babies.[53] It should be remembered that in all surveys researchers tested multiple cells of each individual. In this study they chose one hundred cells per individual for analysis. When discrepancies appeared, a new set of one hundred cells was tested. The procedure highlights the sheer amount of work involved in such surveys.

This first and later screening programs revealed an unexpected large number of chromosomal anomalies, pointing to a potentially explosive public health issue. Findings indicated that about 1 percent of infants showed a chromosome anomaly in their somatic cells. One-quarter of those regarded a sex-chromosome anomaly; one-quarter showed other kinds of trisomies, most of which were "mongols" shown to carry an extra chromosome 21; and the rest showed detectable structural rearrangements. This quite certainly represented an underestimation of the chromosomal anomalies present in the general population because of the difficulties of visualizing such changes with the available methods.[54] At the same time, not all anomalies observed under the microscope corresponded to known clinical symptoms, confounding the distinction between normal and abnormal or pathological and compounding the problem of how to interpret the new findings.

In the following years, and taking full advantage of the peripheral blood technique for karyotyping, the Edinburgh unit embarked on an ever-expanding series of population studies. The newborns were considered representative of the population at large—with the caveat that babies born in hospitals at the time represented a somewhat skewed representation of the general population in regard to socioeconomic status. Other groups the unit studied were defined by a particular condition, such as infertility, or by some other common marker, like being a patient in a psychiatric hospital or having been exposed to a workplace hazard such as radiation or benzene. The aim of these and other epidemiological studies was to identify the causes or, in this case, the genetic basis of a certain condition or behavior or the effects of an exposure. To avoid ascertainment biases, all studies also included control cases. The populations studied at the Edinburgh unit comprised patients in psychiatric institutions, inmates in high-security hospitals, prison populations, students in special schools, infertile or subfertile men and women, women who suffered multiple miscarriages, and individuals exposed to ionizing radiation, such as cancer patients, industrial workers, people involved in the refueling of depleted uranium, and people exposed to radiation leaks at the Windscale accident of 1957, when a fire broke out at Britain's first plutonium factory. Subjects exposed to other toxic substances like aromatic hydrocarbons, herbicides, and pesticides were also screened.

The choice of these populations was driven by the research interests pursued in the unit, which in turn very much depended on the possibilities opened up by the new tools of human chromosome analysis. Next to the study of the hazards from radiation that remained an enduring interest of the Edinburgh group, the genetic basis of mental disability and the biological determinants of sex seemed amenable to chromosome studies. Nevertheless, the study of human heredity had long availed itself of institutions like schools, hospitals, and prisons (next to the military) to provide access to test populations. Chromosome studies followed in that mold, bringing new tools to an old quest.[55]

The number of individuals tested in some of the studies was staggering. Chromosome surveys of the prison population or of patients of mental institutions often included thousands of individuals. For example, a survey of sex chromatin abnormalities in patients of mental hospitals in Scotland conducted by the Edinburgh group in the late 1960s included over thirteen thousand individuals.[56] A chromosome study in penal institutions and approved schools conducted by other researchers of the same group some years later included over twenty-five hundred individuals.[57] Each study made reference to other studies conducted in Scotland or further afield including equally high numbers.[58] In some studies the numbers were not as high but were still exhaustive. For example, the chromosome study conducted at Carstairs, Scotland's high-security hospital, that resulted in the controversial claims about the XYY karyotype, included all 342 patients then held in the institution, except for 27 who "refused to collaborate."[59] Access here, as in many other cases, was regulated through the in-house physician, who provided researchers with blood samples obtained in what was described as "routine medical examination."[60] When researchers were medically trained, they sometimes received permission to do their own sampling.[61]

In the early 1960s, Court Brown submitted a memorandum to the

WHO, one of a series of international organizations that became concerned with the public health aspects of radiation exposure, suggesting various ways in which chromosome studies could be used as biological indicators of radiation damage in large-scale population studies.[62] When, ten years later, the WHO issued a manual on chromosome aberration analysis for the measurement of radiation damage and environmental mutagenic effects, including a vastly expanded array of cytogenetic methods and applications, Karen Buckton from the Edinburgh unit had the leading hand. The plans envisaged "mass screening" of the general public.[63]

In addition to these extended screening projects, small, isolated populations attracted the attention of genetic researchers. In 1966, a team of scientists from Edinburgh descended on the island of Barra, the southernmost island of the Outer Hebrides off the western coast of Scotland, to carry out chromosome studies on all members of the population that were over sixty-five years old. The island was a remote place. The inhabitants had a record of longevity, and extensive genealogical records were available for all members of the community. All this made the island inhabitants in the eyes of the researchers an ideal population to study the effects of aging on chromosomes. Once chromosome banding became available, the island samples and the chromosomes of newborns were compared, and certain differences in the distribution of banding variants were observed.[64]

The investigation of the inhabitants of the island of Barra can be viewed as an extension of the epidemiological studies pursued so energetically at Edinburgh. Yet geographically and culturally isolated populations had long attracted intense interest for the study of human variation and human evolution.[65] In the fractured postwar era, characterized by nuclear concerns and projects of modernization and development, that interest was picked up and fostered by international organizations, most prominently the WHO and the IBP. There certainly were overlaps between the epidemiological and population-based studies of human variation, for instance when epidemiological studies included ethnic differences as a criterion.[66] Yet the use of genetic technologies, including karyotyping, for the study of "racial variations" carried different cultural baggage. The issue had preoccupied chromosome researchers for a long time. The scale of the early studies was rather modest, but their scope not less far reaching.

The Study of Human Variation

In early studies on human chromosomes in the 1910s and 1920s, the possibility was entertained that "whites" and "Negroes" had a different number of chromosomes. Thomas Hunt Morgan, head of the celebrated Fly Room at Columbia University and a staunch anti-eugenicist, suggested this possibility as a way to harmonize the counts presented by Belgian cytologist Hans von Winiwarter, who had been working on tissue of men of European descent, and Michael F. Guyer, a zoologist at the University of Wisconsin who had been using samples from African Americans.[67] Guyer later supported this claim by reporting higher chromosome counts in samples of "two Caucasians."[68]

By the mid-1920s the possibility aired by Morgan was laid to rest, and it was generally accepted that "whites" and "Negroes" as well as "Japanese" and women and men all had the same number of chromosomes (fig. 4.9).[69] The consensus was part of the effort to stabilize the human

FIGURE 4.9. Comparative numbers of chromosomes in humans and other mammals (1925). The chromosome numbers of "whites" and "Negroes" were recorded separately (with the one for "whites" always listed first). The count for both was forty-eight.

Source: Painter, "A Comparative Study," 392, plate 1. Reproduced with permission conveyed through Copyright Clearance Center Inc. of University of Chicago Press Journals.

chromosome count, which was set at forty-eight. Scientists agreed on this point at the height of the eugenic movement and Jim Crow segregation. Indeed, the Texas-based zoologist Theophilus Painter, who was instrumental in closing the debate on the number of human chromosomes, later in his career as interim president of the University of Texas at Austin, became the defendant in the high-profile court case *Sweatt v. Painter* challenging segregation at the school. Painter did not play this role inadvertently. Rather, the previous president, who thought segregation should end, lost the confidence of the trustees, and Painter was seen as a choice who would support the school's segregationist position.[70] Clearly, it was possible to argue for the same number of chromosomes for all people and still support segregation. Nevertheless, the idea that there might be chromosomal differences between various populations was never quite abandoned. When in the mid-1950s the number of human chromosomes was revised from forty-eight to forty-six, the question was again put to the test.

An early challenge to the still-tenuous consensus on the new chromosome count came from Masuo Kodani, a Japanese émigré who worked both in the United States and with the Atomic Bomb Casualty Commission in Japan. Kodani reported counts of forty-six, forty-seven, and forty-eight chromosomes in samples of Japanese males. The higher counts depended on the presence of either a single or a pair of what the author described as small inert "supernumerary" chromosomes. Kodani expected—and later confirmed—that the same three counts also existed in "whites." He suggested that these observations could explain the divergent chromosome numbers reported in the literature. Moreover, his studies indicated that the proportion of the three possible chromosome numbers could vary in different ethnic groups, with a forty-eight-chromosome count being more likely in Japanese people than in "whites."[71]

Kodani was an experienced cytogeneticist. Nevertheless, Ford, an authority in the field who had the opportunity to study Kodani's photographs, quickly decided that a count of twenty-three bivalents (or paired chromosomes) was the more "plausible" interpretation in all preparations.[72] The case seems to have rested there, but in fact Ford did not dismiss Kodani's project altogether. An exhibit on human chromosomes in the living-cell section of the exhibition at the International Science Pavilion at the Brussels World Fair in 1958, for which Ford signed as responsible, explained to the millions of visitors flocking to the fair that "it is now known that 46 is the correct number [of chromosomes in the body cells of human beings], at least in Europeans."[73] In a review article on the status of the field a few years later, Ford encouraged further studies of chromosomal variation, contending that "comparison of the chromosome sets of different ethnic groups immediately suggests itself as the most likely method of revealing polymorphism if it exists. It is unnecessary to stress the interest for anthropology if any form of chromosome polymorphism should be revealed."[74] In a careful study drawing on a large sample, Sajiro Makino from Hokkaido University in Sapporo, on visit at Hsu's laboratory in Texas, confirmed that Japanese people indeed had forty-six chromosomes, as did their Western counterparts, but not before another study suggested that Chinese people also had more than forty-six chromosomes.[75] Remote populations that were geographically, culturally, or reproductively isolated often were the preferred subjects of study for the continued search for possible variations in chromosome number and shape. The following episode, taking place in Ford's laboratory, is telling.

Shortly after Kodani's challenge had been settled, Harnden, then a postdoctoral student in Ford's laboratory at Harwell, developed a new technique for growing a skin biopsy from his own arm to be used for karyotyping.[76] Harnden maintained that he was probably the first person in the world who looked at his own chromosomes. He later recalled the exhilarating experience and the feeling that he was "able to explore areas not reachable by anyone else."[77] One of the first projects he pursued with his new technique was to test the hypothesis that different human populations carried different numbers of chromosomes. If such variation existed, he reasoned, it was most likely that it would occur in populations that had been geographically isolated for a long time.[78] With the help of a colleague in Adelaide, he organized for a skin biopsy of an "assuredly 'full-blooded' aborigine" to be flown to him for testing via the Royal Flying Doctor Service.[79] As he reported, the culture "grew beautifully but the chromosomes were quite normal." Harnden recalled that this was a "disappointment, I suppose, but still fun to do."[80]

Harnden does not seem to have pursued this line of research any further, although he considered his foray into the study of the biology of indigenous people significant enough to mention it in his brief autobiographical essay published in the fiftieth anniversary year of the recount of human chromosomes. Undoubtedly, his study resonated with, and profited from, broader concerns in postwar anthropology and human population genetics that maintained an interest in populations that were geographically, culturally, or reproductively isolated. Many of these populations were already subjects of intense study and exploitation in colonial and postcolonial settings.[81] Harnden took the category of "aborigines" as given, and in effect tested a single sample as if it were representative of a type, neglecting the assumptions that went into constructing such categories.[82]

Others followed in Harnden's tracks. Among cytogeneticists, Hungerford, the co-discoverer of the peripheral blood method that had opened up chromosome analysis to wider use, teamed up with physical anthropologists to collect and study samples of indigenous people in eastern New Guinea, the Todas of southern India, and the Ainu on the island of Hokkaido in Japan.[83] He credited the British anthropologist Nigel A. Barnicot and his associates at University College London with having initiated the systematic search for chromosomal variation in human populations. The London team had compared chromosomes from people in West Africa, Greenland, and Europe, largely drawing on London's cosmopolitan population.[84] As Barnicot's earlier studies also Hungerford's, stretching over several years, showed no recognizable differences. And yet Hungerford remained optimistic that "microscopically visible karyotype variability" could be found, and he encouraged physical anthropologists and chromosome researchers to collaborate to find such variations, as long as "discrete isolates" still existed.[85] Among those who heeded the call was the Italian population geneticist Luca Cavalli-Sforza. In the mid-1960s, while undertaking his extensive population genetic study of the "Babinga Pygmies" in western Africa, he sent blood samples back to Pavia, Italy, for chromosome analysis. Once more the chromosomes were found to be "normal."[86]

We should remember that obtaining the samples for such studies was all but straightforward. Expeditions such as the one conducted by Cavalli-Sforza required extensive preparation. First, pilot studies were undertaken in which conditions on the ground were explored, personal contacts with local administrators and chiefs established, study areas and villages chosen, and transport routes for people and samples tested. If the preliminary investigations were encouraging, then full-scale expeditions with multidisciplinary teams of demographers, anthropologists, geneticists, physicians, and technicians were planned. The studies often required two or more years of presence in the field and follow-up visits. Cavalli-Sforza listed a Jeep, a Land Rover, an electric generator, two refrigerators, a walkie-talkie system, and two lightweight tents with special netting, as well as an ample supply of sterile vacuum test tubes, sterile blood sets, and plastic syringes for blood collection, as essential items for the safety and success of his expedition to West Africa.[87] The other biggest and most important additional cost to avert "disaster" concerned the organization of safe and swift transport of refrigerated blood samples from remote field sites to laboratories in major centers where they could be analyzed.[88] Not explicitly mentioned in this list was the collaboration of the study subjects, on whose participation the investigation crucially depended. Support for Cavalli-Sforza's study and other genetic studies of "vanishing" or "primitive people" came from the WHO and the IBP, both of which included large programs in human heredity.

Genetic Surveys on a World Scale

The WHO program in human heredity developed out of the organization's interest in the public health implications of atomic radiation.[89] Interest in this field started in the mid-1950s, at a time of increasing political and public concerns about global radioactive fallout from hydrogen bomb testing and the simultaneous expansion of programs for "peaceful" and commercial uses of atomic energy.[90] In view of accumulating data on worldwide contamination of radioactive strontium and other elements and their carcinogenic effects, in 1956 the World Health Assembly recognized protection against radiation as a global public health issue and a new area of responsibility for the WHO. The genetic effects of radiation soon emerged as the WHO's main focus.[91] A few years later, officers of the organization affirmed that the WHO's interest in human heredity was "by no means confined to radiation genetics," but rather covered the broad field of human genetics.[92] By 1959 WHO-sponsored projects in human genetics included genetic studies of human populations exposed to high natural radiation; a study on the distribution, treatment, and prevention of heritable blood diseases, including thalassemias and other forms of anemias; a large-scale survey of congenital malformations in newborn babies in twenty countries around the world, coordinated by Alan Stevenson, director of the Population Genetics Research Unit in Oxford; and a study of populations of "particular genetic interest," including especially "isolated" and "socially primitive groups," spearheaded by the American geneticist James Neel. Neel had made his name as chief scientist of the Atomic Bomb Casualty Commission and its genetic project in Japan and was building up a Department of Human Genetics at the University of Michigan in Ann Arbor, one of the first of its kind in the United States.[93] Here and elsewhere, "primitive" groups were defined as communities that "still obtain food by means that were prevalent in the early phase of human history, such as hunting and food gathering, nomadic pastoralism, and digging-stick-and-hoe type of agriculture."[94] While the comparative birth study was expected to provide important insights into the distributions of various mutations and malformations and the respective role of genes and environment in disease formation, the study of "unusual" populations was meant to provide information on natural selection and the role of biological and social factors in mating patterns, fertility, morbidity, and life expectancy, among other things.[95] All projects required the coordination of international teams of researchers and the management of large data sets. The WHO offered legitimization, infrastructural support, and limited funds.[96]

Neel, whose project is of most interest to us here, soon embarked on a study of the Xavante Indians in Brazil that he hoped would serve as a model for studies to be carried out in other parts of the world.[97] He also threw himself into the organization of what he presented to the WHO officials as a "milestone" conference, named "Population Genetics of Primitive People."[98] The conference agenda included a discussion of the concepts of population genetics, of the rationale and need for studies of "primitive populations," of standard protocols for such studies, and a survey of populations suitable for study. Anthropologists, demographers, and geneticists were to be among the participants. Whenever possible, they were to be recruited from countries where the target populations lived. The final aim of the meeting was to produce a "standard reference manual" that could serve as a basis for a series of studies.[99]

The project on the study of "primitive" people at the WHO proceeded despite strong reservations by some of its officials. For instance, when asked to approve a grant to Cavalli-Sforza's study on the Babinga under the project title, the WHO's assistant director general Dr. John Karefa-Smart, of Sierra Leone, expressed his strong objection against this kind of "anthropological excursion" in view of other "urgent priorities" in Africa and elsewhere. Other officers agreed that the study would help genetical research rather than public health issues in Africa but nevertheless regarded it as important to have "data on at least one interesting group in Africa" in the broader context of the project.[100] In response to all objections, the promoters of this and other such studies always argued that the "primitive populations" were quickly "vanishing" because of advancing civilization and that studying them was urgent to gain insights into the deep history of humanity and to understand the evolutionary pressures and reproductive patterns that had determined the "genetic attributes of civilized man."[101]

The trope of vanishing populations was first introduced by the Swiss zoologist, ethnographer, and nature conservancy advocate Paul Sarasin in the 1910s. Whereas in an evolutionary framework the extinction of "primitive people" would seem inevitable, Sarasin had argued that "primitive people" (or *Naturvölker*) needed to be protected because of their scientific value.[102] As Neel and others argued, studying the genetic structure of these populations was all the more urgent at a time when human populations around the world were exposed to new genetic risks. Understanding the impact of these changes required extended knowledge of the selective pressures to which humans were exposed during their history. Hence, studying the past as preserved in the few surviving populations still untouched by modern civilization provided an essential key to understanding the present and future of humankind.[103]

In addition to the scientific value of the "salvage" operation, the researchers stressed the benefits the studies brought to the local populations.[104] Every study subject received medical and dental treatment. Moreover, the health and demographic assessment helped local administrators identify the problems that required most urgent attention. The WHO guidelines for research on human population genetics also highlighted the ethical obligations of the researchers toward the people under study, as well as the needs to seek consent, to respect the "privacy and dignity" of the individuals, to guarantee their "comfort," to maintain their anonymity, and to respect the "cultural integrity" of every group.[105] The need to draw up these guidelines points to the precarious conditions of genetic fieldwork.

The population studies combined anthropological, genetic, medical, and demographic approaches, all of which were seen as instrumental to a "comprehensive study" of the genetic structure of a population. The more strictly genetic studies at the time were mainly based on the analysis of developmental malformations and on the study of variants and frequencies of hemoglobins or other serum proteins. They also included pedigree analyses and consanguinity studies. Yet karyotyping increasingly became part of the tool kit of human population studies.[106] Already in 1960, the WHO offered an international training course in chromosome techniques. Hosted by the Anatomical Institute in Basel and gathering together twenty participants from nineteen European and neighboring countries, who had been selected from a much broader pool of applicants channeled through local governments, the course was seen as an effective way to disseminate the new techniques. WHO officers anticipated that further such courses would be arranged in other regions.[107] The study of chromosomes also featured prominently in WHO publications that dealt with the relations of genetics and public health. For instance, the 1966 summer issue of its widely circulated glossy magazine *World Health* was dedicated to the theme "Genetics and Your Health."[108] In a series of sharp black-and-white photographs by the documentary photographer Paul Almasy, Lejeune, pictured in his Paris laboratory, demonstrated the various steps involved in constructing a karyotype (see fig. 2.8).[109] More generally, an important role of the WHO was promoting the standardization of procedures and terminology and setting up international reference centers for specialized identification of genetic variants. It also supported the development of computer technologies to help with various aspects of genetic research, including statistical analysis, modeling techniques, data linking, and storage. For instance, in the mid-1960s the WHO supported the Human Genetics Computer Project at the University of Aberdeen, where the British population geneticist Anthony W. F. Edwards, after having spent several years working with Cavalli-Sforza first in Pavia and then at Stanford, developed computer programs for constructing pedigrees from genetic information and simulating gene flow through a given pedigree. This was useful for calculating consanguinity and gene linkage, and the hope was that the methods could be expanded to study large complex populations.[110] WHO also made its own computer facilities available to researchers. From 1965, the WHO participated in the human heredity component of the IBP.

Discussions for an IBP, modeled on the highly successful International Geophysical Year of 1957–1958, started in 1959. Italian geneticist Giuseppe Montalenti, then president of the International Union of Biological Sciences, who was involved in the discussion, proposed including a human genetics program. He was particularly keen to map gene frequencies of "isolated populations."[111] When the British embryologist Conrad Waddington took over the presidency of the International Union, and hence the planning for the IBP, he strongly advised against the human genetics part to avoid politically controversial issues, notably the question of racial differences, which lacked an adequate scientific understanding. It was nonetheless considered important to include a "human part" in a program oriented toward the conservation of biological resources and environmental change. This happened under the rubric of the "human adaptability program," which took the adaptation of humans to changing environments as its topic. Exactly what form this project should take was not quite clear initially.[112]

In 1964, a meeting with the title "The Biology of Populations of Anthropological Importance" was convened under the auspices of the IBP at Burg Wartenstein near Vienna, the European conference center of the Wenner-Gren Foundation for Anthropological Research. Its aim was to review the present state of knowledge in the general field of "human adaptability."[113] The meeting revealed "a staggering state of ignorance about the human race."[114] Genetic surveys were suggested as a way to fill that vacuum. It was noted that genetic studies of "primitive people" or "vanishing populations" were already going on under the auspices of the WHO. Neel, together with his Brazilian colleague Francisco M. Salzano, presented "A Prospectus for Genetic Studies on the American Indians," based on their study initiated under the aegis of the WHO that they hoped could serve as a model for other such studies.[115] In their talk they succinctly summarized the three main research interests that guided their investigation: the question of the genetic divergence of human populations and the tempo of human evolution, the study of biological parameters of people in the "pre-Columbian state," and the study of disease patterns "when these primitive groups make the transition from a near-Stone Age to an Atomic Age existence."[116]

The meeting at Burg Wartenstein was regarded as an important planning meeting for the IBP. A last-ditch plea by noted social anthropologist Margaret Mead to substitute the human adaptability proposal with a program based on the social sciences was not successful.[117] Although what we might today consider "racial" questions were not absent in the human adaptability program, the final report drew a clear distinction between the "old-fashioned and static subject of physical anthropology" and an "ecologically and genetically based" approach to human population studies that informed the investigations undertaken as part of the IBP.[118] The new approach was also captured under the label of "human biology," a term in circulation since the 1930s but gaining new currency at the time. An introductory volume with that title, published under the auspices of the IBP, was hailed as "the first to treat it as a subject of study in its own right."[119]

Data collection was vast. *A Guide to Field Methods*, compiled by the convener, London-based human biologist Joseph S. Weiner, in collaboration with the scientific coordinator of the human adaptability program, provided detailed descriptions of about fifty procedures that were presented as "essential to the biological study of human populations."[120] These included blood collecting and a gamut of physiological tests, as well as techniques of medical examination and demographic assessment. Cytogenetics found its place alongside a variety of other genetic tests, such as blood typing, color vision testing, skin color measurements, and the analysis of fingerprints and palm prints. The collection of data on the frequency of chromosome anomalies in different populations was regarded as of "considerable value." The respective techniques were considered "simple" and easily adaptable to field conditions. Buccal smears could offer insights into sex-chromosome anomalies "within an hour or so of collection," provided that a microscope and simple laboratory facilities were available. Blood smears for chromosome analysis were perhaps even easier to prepare by the "average technician." Analysis, however, required skilled personnel and had to be performed in a collaborating cytogenetic laboratory. Skin biopsies, when collected in sterile culture medium and kept cool, had the advantage of resisting transport periods of more than seven days (the maximum allowable for blood samples). They were thus "the method of choice in the more remote areas." Results with necessary identifications were to be recorded on a standardized data-collecting sheet for further evaluation.[121]

Ten years later, the final report of the Human Adaptability Program listed several cytogenetics projects or projects with a cytogenetic component. The list included a study of chromosome mosaicism in carriers of Down syndrome as well as a series of other studies of clinical populations in Brazil;[122] studies of the "Eskimos," conducted over a five-year period by Canadian researchers;[123] Cavalli-Sforza's genetic study of the Babinga; studies of several indigenous South African populations, including the "Hottentots, Bushmen, Damara and the Rehoboth Bastards";[124] and of the islanders of Tristan da Cunha.[125] The thinly populated island in the Atlantic was presented as the "most remote inhabited location on Earth." In 1961, following a volcano eruption on the island, the whole population of 264 people was evacuated to Britain. In the two years the people spent in Britain, they became the subject of extensive medical and scientific investigations. Over fifty researchers participated in the studies. The results of the cytogenetic studies were published in a paper in *Nature* under the title "Chromosome Investigations of a Small Isolated Human Population." Despite widespread congenital malformations present in the population, no chromosome anomaly was found.[126]

Neel also included a cytogenetic study in his fieldwork among the Yanomama. The goal was to establish a baseline for chromosome images of "a truly primitive population, with no known exposure to medical radiation, food preservatives, pesticides etc."[127] Blood samples were collected from forty-nine Yanomama Indians living in two villages near the Venezuelan-Brazilian border. The samples were immediately chilled in a portable refrigerator and flown to a temporary cytogenetics laboratory established at the Venezuelan Institute of Scientific Investigation near Caracas. Expedition members supplied control samples. Cultures were grown and slides prepared that were sent to Ann Arbor for analysis. To their surprise, researchers found a significantly larger percentage of cells with severely damaged chromosomes in the "Indian samples" than in the controls and even in controls of Japanese survivors of the atomic bomb. The researchers speculated that measles immunization, a virus infection, or toxic plants could have induced the damage. Follow-up studies did not show the same phenomenon, throwing the first findings into doubt. Only much later did it emerge that the shower of abnormal cells observed in the first study was a phenomenon that occurred periodically in people all around the world. The causes were unclear, but the rearrangements could well be the starting point for the formation of a cancer. In retrospect, Neel considered the observation of the "rogue cells" that he also described as the "cytogenetic surprise" as "potentially the most exciting discovery to come out of the 'Indian Program.'"[128]

The Edinburgh group was not involved in the studies conducted under the IBP, although the field guide prominently referred to data on the frequency of chromosome anomalies in the general population published by the group. The recognition of the expertise of the group can also be gauged by the fact that, in the early 1970s, the unit at Edinburgh was designated as the WHO International Reference Center for Chromosome Aberration Evaluation in Populations.

In the same year the final IBP report was published, Jacobs—who by that time had moved to a new position at the University of Hawai'i School of Medicine but was still in close contact with her Edinburgh colleagues—published a review article, "Human Chromosome Heteromorphisms," in which she reviewed the potential of karyotyping for studies of human variation.[129] In a later biographical essay, Jacobs pointed out the "fortunate" circumstance that Hawaii offered an ethnically highly diverse study population that comprised people of Chinese, Japanese, Filipino, Hawaiian, and Caucasian ancestry. This enabled her to study the different prevalence of various conditions among the ethnic groups.[130] The same circumstance may explain her interest in chromosome heteromorphisms in various population groups.

Jacobs started her article stating that since the "rebirth" of human cytogenetics in the late 1950s, it had been recognized that morphological differences existed on chromosomes of single individuals. These variations did not have an obvious clinical effect but were inherited in a Mendelian fashion. Banding techniques introduced in the early 1970s had made more such differences visible. The observed heterochromatic bands differed in size, position, staining intensity, or a combination of these variables.[131] Yet the distinctions were at the limits of resolution of the light microscope and were extremely difficult to quantify. For this reason, results between laboratories could not easily be compared, and differences attributed to ethnic background of the populations studied "cannot be taken seriously at this time."[132] However, the situation changed if the studies were done by the same laboratory and scored blindly.

Under the heading of "racial variation," Jacobs then listed various reported differences in chromosome heterochromatin. The first such case, published in the mid-1960s, described variations in the length of the Y chromosome in "Japanese," "Jews," "non-Jewish Caucasians," "Negroes," and "American Indians" (with Japanese having the longest and American Indians the shortest Y chromosome). Further studies reported similar variations in the length of the Y chromosome in Australian Aborigines (shorter than in Caucasians) and between two Indian populations, the Rajputs and the Punjabis.[133] The variation in length was found to rely entirely on differences in length of the heterochromatic distal segment, the remaining chromosome being identical in all cases (fig. 4.10). The focus on the Y chromosome was at least partly due to a clear fluorescent band on the long (gene-poor) arm of the chromosome that reliably showed up and was easy to measure. There was no evidence that the heritable variation in length was associated with any observable effect on development or behavior. Exactly for this reason it was a useful marker. Jacobs moved on to review various studies showing heterochromatic differences between "Negroes" and "Caucasians" (with African

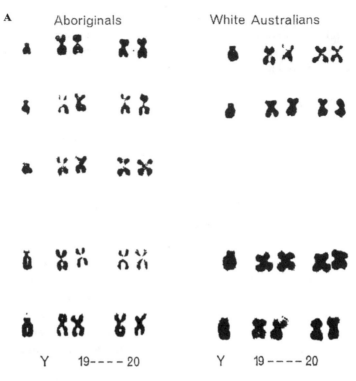

A   Aboriginals   White Australians

FIGURE 4.10. Variation in the length of the Y chromosomes in "Aboriginal" and "white Australian" groups. (A) Partial karyotype of chromosomes 19 and 20 and the Y chromosomes of five "Aborigines" and four "white Australians" showing the shortest and longest Y chromosomes in each group; (B) diagrammatic representation of the variation in length of the fluorescent segment in the Y chromosome of "Aborigines" and "white Australians."

Source: Angell, "The Chromosomes of Australian Aborigines," 106, fig. 1, and 107, fig. 3. Reproduced with permission of Aboriginal Studies Press.

Americans showing more such bands). The review of all these cases once more highlights the number of population studies that were undertaken. Jacobs concluded that there was "considerable variation in chromosome heteromorphisms among different racial groups" and that further investigation of the phenomenon would provide valuable information "on the origin, migration and kinship" of these groups.[134] In addition, heterochromatin was useful for identifying and tracking chromosomes (for instance, to trace the origin of the extra chromosome in Down syndrome), cells (to establish paternity), and individuals, and it was playing an important role in linkage studies of particular genes. For instance, the first assignment of a gene to an autosome—that is, not an X or Y

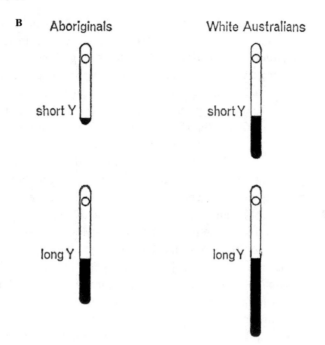

FIGURE 4.10. (*continued*)

chromosome—in humans, that of the Duffy blood group to chromosome 1, was made by showing that the Duffy blood group gene and a large heteromorphic region on chromosome 1 were linked. Jacobs concluded that when "objective methods of mensuration" would become available, "heteromorphisms will take their place alongside conventional blood group and enzyme polymorphisms as tool in formal and population cytogenetics."[135] Other researchers echoed her view.[136]

Old Wine in New Bottles?

Following the use of human karyotyping has brought into focus a broad range of genetic studies of human populations carried out between the late 1950s and the mid-1970s that are rarely mentioned in postwar histories of genetics. We already saw karyotyping playing a key role in efforts to investigate and monitor the effects of radiation in an era of vastly expanded military, industrial, and medical uses of atomic energy. With rising concerns about the hazards of nuclear and chemical mutagens,

karyotyping was also presented as an effective tool for "population surveillance."[137] Human chromosome diseases, first described in the late
1950s, were regarded as a new public health issue, giving rise to multiple screening programs to study and control the problem. Karyotyping piggybacked on ongoing medico-anthropological-genetic studies of
indigenous people and other populations around the world that were
sponsored by the WHO and the IBP. Cytogeneticists kept up the expectation that chromosome analysis would eventually develop into a more
powerful tool for the study of human variation and for gene mapping
than blood groups and serum proteins—a promise eventually fulfilled
by somatic cell genetics and genomics. Meanwhile cytogeneticists participated in the sampling and collecting of genetic data on a large scale,
in the creation of central resources such as registries and tissue repos-
itories, and in efforts to harness modern computers to help with the
scale and complexities of their projects. The WHO facilitated international collaboration by hosting meetings, encouraging the standardization of techniques and nomenclatures, publishing technical manuals and
reports, and setting up a series of reference laboratories. We also saw
the WHO embracing genetics as part of its global public health agenda
in the Cold War era. Propelled by state funding, radiation and public
health concerns, the development of common standards, and international cooperation, these efforts helped make space for and expand the
reach of human genetics in the scientific, public, and political discourse.
Chromosomes emerged as the central objects that pushed forward and
reflected these changes.

Since the 1960s ethical guidelines have changed, and many aspects
of the genetic population studies of earlier decades have become problematic or seem obsolete. Whole genetic data or sample collections were
thrown into limbo. Yet in the post-genomic era, high hopes are again
pinned on human population studies. Population geneticists using powerful new-generation sequencing techniques and statistical software pack-
ages promise to offer new insights into human evolution and history and
to provide the basis for a new genomic medicine tailored to individuals.
The strongly contested Human Genome Diversity Project, initiated by
Cavalli-Sforza in the early 1990s, clearly followed in the footsteps of the
"Study of Primitive People" supported by the WHO and the IBP in the
1960s. Employing some of the same rhetoric, Cavalli-Sforza and his colleagues, in their call for action, pointed to the "vanishing opportunity"
to collect blood samples from quickly disappearing "isolated human

populations" around the world who held the key to the study of human diversity. They also called on the WHO, as well as the Human Genome Organization and other institutions, to support the urgent international effort.[138] The project encountered resistance and eventually floundered. Among those who contested the project were the indigenous populations included in the study who protested their description as vanishing isolates of historical interest and opposed the scientific exploitation of their genetic heritage.[139] A significant outcome of these debates has been the elaboration of new protocols for the use of existing and new human tissue collections that better respect the sensibilities, rights, and needs of indigenous people involved in the studies.[140] Nevertheless, the HapMap Project, which catalogs common genetic variants and their distribution, and more recently, the 1000 Genomes Project launched by an international consortium, have pursued the project of recording human genetic diversity. Similarly, the UK Biobank, started in 2007, collects genetic, medical, and lifestyle information for five hundred thousand UK citizens, and the Deciphering Developmental Disorders project, led by the Sanger Institute at Hinxton, which studies copy variations in one hundred thousand children, can be viewed as the "logical follow-ups" of the Edinburgh Registry of Abnormal Karyotypes and the newborn-screening programs.[141] The BabySeq Project, launched in 2015 and funded by the US National Institute of Child Health and Human Development and the National Human Genome Research Institute, also explores the integration of sequencing data into the care of newborns. Racial and ethnic variations are regularly recorded in many of these studies, especially in the United States, where this was federally mandated.[142] Only by taking into account the whole spectrum of genetic projects—from chromosome surveys to studies of molecular structure and function that were pursued side by side—is it possible to start understanding the contours and scope of postwar genetics, including especially human genetics, and its more recent genomic developments. The sometimes-fraught relations between microscope-based studies of chromosomes and molecular approaches to heredity are the subject of the last chapter.

CHAPTER FIVE

Of Chromosomes and DNA

In the late 1970s, considering the future of cell genetics, Hsu, a doyen of the field, mused that "probably the most pressing problem in chromosome research is the understanding of the molecular architecture of the chromosome."[1] A few years earlier, Francis Crick, of DNA fame, had declared the structure of the chromosomes of higher organisms to be "probably the major unresolved problem in biology today."[2] Around the same time, James Watson organized the Cold Spring Harbor Symposium on Chromosome Structure and Function, at which Crick and other leading molecular biologists presented papers. In the foreword to the conference proceedings, Watson confidently predicted that "the essential structural features of chromosomes may be resolved over the next decade."[3] Hsu was glad to note that the "glaring gap" that had existed until recently between the biologists studying the molecular components of chromosomes and those studying the chromosomes themselves was beginning to narrow. This, he enthused, may indeed "realize the dream of all cytologists from the last century to the present day: knowledge of the structure and function of chromosomes."[4]

Hsu's position on this issue may well have been unusual among cell geneticists. Hsu himself considered it "sad that many cytogeneticists have not even made an attempt to learn a little about molecular biology." If they had, he reasoned, they would discover the pleasure of "deepening their field of inquiry."[5] Moreover, Hsu's remarks prompt a series of questions: How can we account, historically and epistemically, for both the "gap" between molecular biologists and chromosome researchers and the supposed narrowing of that gap between the two research communities? What exactly was involved in the "molecularization" of cytogenetics implied by Hsu, and what can we learn from it in regard to the history of chromosomes and DNA and their respective place in the history of postwar heredity?[6]

In his *Image and Logic* the historian and philosopher of science Peter Galison presented the postwar history of particle physics in terms of two competing instrumental and epistemic traditions—one bound to imaging technologies, the other to analytic techniques of calculation. According to Galison, these two traditions came together with the advent of electronically produced images. At this point, imaging was dissolved into quantifiable pixels, a transformation that allowed for a "fusion" of the techniques of imaging and calculation.[7]

A similar account could hold for the history of postwar genetics with the microscope-based tradition of cytogenetics running on a separate track from the analytical tradition of molecular biology.[8] Not unlike the case of particle physics, cytogenetics and molecular biology are sometimes presented as eventually "fusing" with the advent of fluorescent in situ hybridization, or FISH, techniques.[9] The procedure uses fluorescent probes, produced by recombinant DNA techniques, to detect and localize specific sites on the chromosomes. The result can be viewed under a fluorescence microscope and in practice allows for a molecular resolution of the chromosome image. First developed in the early 1990s, this technique is often regarded as marking the beginning of molecular cytogenetics.[10] Yet such a label on its own does not give away enough about the dynamics and complexities of the changes involved. Indeed, some cytogeneticists, including Hsu, writing even before the advent of FISH, predicted a glorious future for their discipline. More commonly, though, cytogenetics is seen at best as a prelude to the triumphant development of molecular genetics.[11] In their foreword to the proceedings of the eleventh International Chromosome Conference in 1992, the editors, two cytogeneticists, hinted at the conflicted nature of the changes in course: "A few years ago it may have seemed to some that chromosome studies were being superseded by molecular biology, but the molecular biologists have now realized that they need to know about chromosomes, and indeed an important, if ill-defined discipline of 'molecular cytogenetics' has grown up in recent years. We are pleased that in planning this Conference and this book, so much of the work presented is at the interface between cytogenetics and molecular biology. This will surely continue in the future, as boundaries between disciplines are largely artificial, and each has much to learn from the others."[12]

This chapter aims to make space for an interconnected history of microscope- and test-tube-based approaches to chromosomes and DNA by exploring the multiple intersections, dependencies, and continuities of the two research traditions. In particular, it pushes back against the perception that chromosome research was just "old-fashioned" biology that was eventually superseded by molecular approaches.

In the formative years of the new discipline, molecular biologists often celebrated their distinction from "old time" biologists. Committed to a molecular approach to explaining all biological processes, they held a certain disdain for any approach that engaged other levels of analysis. Cytogeneticists peering down the microscope certainly fit into this category. When Crick, who started his biological career under the guidance of the cell biologist Arthur Hughes, was asked about his teacher, he summed up his response: "Well, you see, he was a microscopist."[13] Crick's biographer suggested that, coming from Crick, this was a rather "damning remark."[14] Crick was certainly not alone in pitting "old" and "new" biology against each other. Historians have followed suit, giving much more attention to molecular research practices in post–World War II biology than to supposedly outdated observational research traditions.[15]

Yet Crick, never one to be dogmatic, also held that biological systems were made up of "a hierarchy of levels of organization, the 'wholes' of one level being the parts of the next." Starting the analysis from the bottom or molecular level was not always the best tactic. Rather, Crick believed that a "simultaneous attack at more than one level will in the long run pay off better than an attack at a single level."[16]

Molecular biologists themselves often followed this intuition. For instance, when in the mid-1960s, Sydney Brenner, Crick's closest collaborator for over twenty years, approached the problem of development with the aim of bringing molecular precision to it, he started off by tracing cell lineages and providing a full nerve cell connectivity diagram of his chosen organism, the nematode *C. elegans*. These approaches were well known to "classical embryologists."[17] Echoing Crick, Brenner remarked: "Biological systems encompass many levels and there are many approaches which are valid in their own right without having to be molecular."[18]

Following these more conciliatory tones, and looking more closely at the molecular biologists' concern with the structure and function of chromosomes and the cytologists' focus on gene mapping, a practice now often associated with genomic science, it is perhaps possible to better understand the differences and the intersections of the two research traditions that were flourishing at the same time. Here, the spotlight is also on the changing visual practices employed to study chromosomes and molecules. In so doing, the analysis expands more decidedly from the 1960s into the 1970s and well beyond. Yet as is so often the case, it is necessary to take a step back before we can move forward.

The Structure and Function of Chromosomes

In the mid-twentieth century the study of chromosomes was not the exclusive province of cytogeneticists. In addition to continuing efforts to improve preparation techniques and study chromosomes under the microscope, chromosomes were subjected to a battery of other techniques for assessing their biochemical and physicochemical properties. Biophysicists in particular were interested in applying physical methods to study the structure of chromosomes. John Randall, who built up a large Biophysics Laboratory at King's College London just after the war, proposed focusing research on chromosomes. Maurice Wilkins, joining him in this venture, applied X-ray crystallography to study the packaging of chromosomes in sperm cells before applying the same technique to study the structure of extracted DNA fibers. The rest, as it is said, is history.

The elucidation of the helical structure of DNA, with its two complementary strands, focused attention on the molecular mechanism of replication and the complex mechanism by which the genetic information is translated into proteins. Biochemists and physical chemists besides newly styled molecular biologists entered the fray. Yet the question of how DNA was packed up in chromosomes remained unresolved, and the advances in preparation techniques for studying chromosomes under the microscope in the mid-1950s did not pass unnoticed.

Hugh Huxley, working first at Cambridge and then in Randall's unit at King's College London, used electron microscopy to study the fine structure and mechanism of muscle fibers. Teaming up with Barnicot from the Department of Anthropology at King's College London, who had a keen interest in the comparative study of human chromosomes, he set out to apply the same technique to the threadlike genetic structures in the cell nucleus. Before joining forces, Barnicot had coauthored a couple of articles on the electron microscopy of human hair pigments—a subject of interest for the anthropological study of human populations—whereas Huxley had published articles on the fixation and staining of nucleic acids for electron microscopy.[19] For their common project, the two researchers cultured cells from skin biopsies, a technique that had only just become available for chromosome analysis. They followed most of the established steps for chromosome preparations but omitted using colchicine to keep chromosomes in a more elongated condition. After fixing the cells, they transferred them to an electron-microscope grid, stained them, washed and dried them, and examined the preparations. In their investigations they paid special attention to the small chromosomes 21, 22, and the Y chromosome, as well as to the centromere region of the larger chromosomes. Prompted by the publications from Paris and Edinburgh describing the presence of an extra chromosome in patients with Down syndrome, Barnicot and Huxley also included probes from a person with Down that they received from Penrose and his colleague J. R. Ellis at the Galton Laboratory. On the basis of their observations, they suggested that the trisomy concerned chromosome 22 rather than 21—as would later prove correct.[20] However, on the whole, the procedure did not give away many more details than could be seen under the light microscope. Barnicot and Huxley were disappointed. Nevertheless, their work inspired other researchers to try their own hand at using electron microscopy to reveal finer details of the structure of chromosomes.[21]

In the following years, the move of molecular biologists to more complex model organisms to study problems of development and nerve function intensified the interest in questions of chromosome structure. In contrast to phages, viruses, and bacteria (also known as prokaryotes), multicellular organisms (or eukaryotes) have a nucleus that contains more than one linear chromatin fiber. The packaging of DNA in eukaryote chromosomes is different and considerably more complex from that in prokaryotes and undergoes changes during the cell cycle.

By the end of the decade, enough data on chromosome structure in higher organisms had accumulated for Crick to attempt to synthesize it and propose a model. The style of his article, a brief note in *Nature*, is reminiscent of Watson and Crick's famous note in the same journal nearly twenty years earlier, and it appears that Crick felt he was up to a similar feat.[22] The article boldly proposed "a general model for the structure of chromosomes of higher organisms." It suggested that intricately folded regions containing the large regulating portions of the DNA molecule were followed by extended stretches containing the shorter coding regions (fig. 5.1). Extended and folded regions corresponded to the interband and band regions that were visible in the giant chromosomes

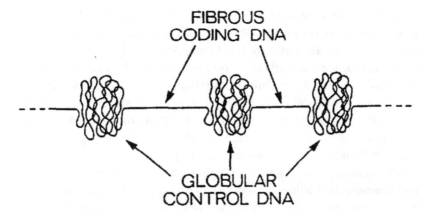

FIGURE 5.1. Crick's "extremely schematic drawing" of the proposed structure of chromatin. Source: Crick, "General Model," 26, fig. 1. Reproduced with permission of Springer Nature.

of the fruit fly. Crick went into much further detail regarding the pairing and unpairing of DNA, the role of proteins attached to the DNA, and the mechanism by which transcription is regulated. He used diagrammatic sketches accompanied by page-long captions to explain the various aspects of the model, which he described as "speculative" but "logically coherent" and "compatible with a large amount of experimental data obtained using very different techniques." In a more modest mode, he suggested that it "raises at least as many questions as it attempts to answer."[23]

Yet the chromosome structure was not to be revealed "with a flourish" like that of DNA.[24] Others quickly found flaws with Crick's model. Nevertheless, the paper did stimulate Roger Kornberg, an American postdoctoral student who would come to the Laboratory of Molecular Biology in Cambridge in 1972, to pick up the problem. He was looking for a "messy subject" to work on and the problem of chromosome structure appealed to him straightaway.[25] After trying unsuccessfully to gain clearer X-ray diffraction pictures of chromatin fibers, he reverted to his biochemical skills. Using calf thymus, he isolated the four different histone molecules that were known to exist and studied their binding properties with DNA. The decisive insight came when he realized that the DNA double helix would wind around a histone core rather than the other way round. Such core units, or nucleosomes, followed each other like beads on a string—not unlike in Crick's model, only in Kornberg's

model the beads contained the coding regions. Being located on the sur-
face of the beads, the coding DNA was directly accessible for transcription.[26] The bead-like structure had been confirmed by electron microscopy and biochemical methods before, but Kornberg's model added new
biochemical and structural information (fig. 5.2). Crick conceded quickly
that Kornberg's model fit the available data better, and he joined Kornberg and others in the laboratory to work out the further details of the
structure, using X-ray crystallography, electron microscopy, and other
methods. Other groups also worked on the problem.[27]

Cytogenetic observations of banding patterns and their physiological functions had helped molecular biologists develop their molecular
model of chromosome structure. Can we expect that, in turn, the model
had a direct impact on the way cytogeneticists looked at the key object
of their endeavors and that it brought chromosome researchers and molecular biologists closer together? Like molecular biologists, cytogeneticists were intrigued by the nature of the bands. The British medical geneticist John Edwards, who had described trisomy 18, also known
as Edwards syndrome, defined it as "the biggest mystery in cytogenet-

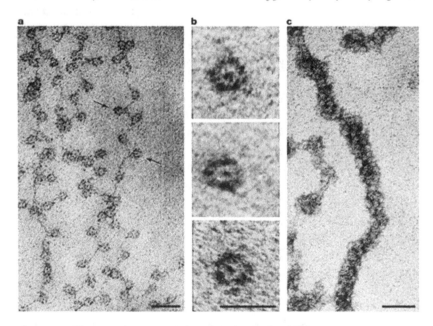

FIGURE 5.2. Electron micrograph of chromatin showing bead-like structure.
Source: Olins and Olins, "Chromatin History," 811, fig. 3. Reproduced with permission of Springer
Nature.

ics" at the time.[28] Picking up on molecular studies, some cytogeneticists intensely studied the dynamics of the banding patterns and their connections to DNA replication.[29] Yet rather than in the general structure and function of chromosomes, cytogeneticists were above all interested in distinguishing individual chromosomes and chromosome variants.[30] Kornberg's model did not speak to these issues and so was not directly relevant to most cytogeneticists. In contrast, the work of Kornberg's team indicated that molecular tools had developed to a stage that they could be used to successfully tackle large and complex cell structures like chromosomes. This emboldened molecular biologists to claim the superiority of their tools. Crick, invited to comment on the proceedings of the International Chromosome Conference he attended in 1977, declared himself surprised about "how little was said about the many promising recent developments in molecular biology," including insights into the molecular structure of chromosomes, genetic engineering, and DNA sequencing. He threw down the gauntlet, declaring: "I feel that chromosome workers will ignore the coming advances in molecular biology at their peril. It's not enough, in order to understand the Book of Nature, to turn over the pages looking at the pictures. Painful though it may be, it will also be necessary to learn to read the text. Only with the assistance of molecular biology will this be possible."[31] Cytogeneticists resented the attack. Without denying the usefulness of a molecular approach, they defended the view that a "radical reductionism" would produce a "distorted view of chromosome structure and function."[32]

Molecular technologies were to gain increasing importance in the cytogeneticists' quest of mapping human genes. Here the interactions between cytogeneticists and molecular biologists became more direct and the stakes higher.

Mapping Genes

Mapping stood at the beginning of the chromosomal theory of heredity. Robert Kohler, in his book on Morgan's fly group, has shown in detail how the move from organizing mutants according to the affected organ systems to describing mutants through linkage mapping introduced a radical shift in the research aims and culture of the group.[33] Linkage mapping was based on large-scale systematic crossing experiments. Genes transmitted together (or "linked") were considered to lie on the same chromosome while the segregation of genes usually transmitted together was explained through the occurrence of rare "crossing over" (or recombination) events between homologous chromosomes. The lower the recombination rate, the closer the genes were expected to lie together. The establishment of four linkage groups of genes, corresponding to the four chromosomes of the fly, and the additive distance between genes stabilized the notion that chromosomes contained genes and that these were lined up on chromosomes like beads on a string. The discovery of giant chromosomes in the salivary glands of the fly gave the mapping project of the "fly people" further momentum. Of special help was that the giant chromosomes, when stained with orcein, showed a banded pattern. Whereas recombination rates were mathematical entities, in the banded giant chromosomes, variations associated with changes in morphology or behavior could be seen under the microscope and physically mapped. As Kohler has pointed out, the *Drosophila* mapping culture rested on a set of distinct literary, technical, epistemic, and social practices. Importantly, fly workers were held together by a "moral economy" based on sharing research tools and results. The work of the fly group established *Drosophila* as a model organism and genetic maps, constructed in a cooperative manner, as the keystone of genetic research.[34]

Other "mapping cultures" also played a role in the formation of the discipline.[35] In the *Drosophila* community itself, geographical mapping was deployed to trace the evolutionary development of fruit flies in the field.

Mapping played an equally important role in the establishment of molecular genetics. In the 1950s and 1960s molecular biologists embraced linkage mapping to study the fine structure of genes and to understand how genes were linked to proteins. Seymour Benzer's fine-scale mutation map of a chromosomal region of bacteriophage T4, one of molecular biologists' preferred model organisms, was hailed as a milestone in understanding the molecular structure of genes and the relations of DNA and proteins. The map was based on classical linkage studies performed on thousands of crosses with T4 phages that carried different mutations in the same gene.[36] The resolution of Benzer's map was on the level of single nucleotides, although the actual DNA sequence remained unresolved. When moving to the study of more complex organisms in the mid-1960s, molecular biologists continued to rely on genetic linkage studies to sort out mutants and the genetic basis of development and behavior.[37] Yet the mapping of human chromosomes remained the province of a growing number of human and medical geneticists who approached the question with classical genetic linkage studies and cytogenetic tools long before molecular biologists joined the bandwagon. The debates surrounding the entry of molecular biologists into the human mapping arena provide fertile ground for the quest of the historical relations between microscope-based and molecular approaches in genetics.

The first mapping of a gene to a human chromosome was in 1911, when Edmund Wilson, after establishing the X and Y chromosomes as the sex chromosomes, suggested that the gene for color blindness must be on the X chromosome, given the sex-specific inheritance pattern of the condition. This was before the fly group started drawing up chromosome maps for their model organism. Through linkage studies of large family pedigrees, a few more genes responsible for the inheritance of diseases, including hemophilia, were located on the X chromosome, resulting in a "provisional map" of the chromosome.[38] In the early 1930s, the idea of using the mass of data accumulated around blood groups as markers for establishing linkage with other genes provided new impetus for the mapping of human genes. Lancelot Hogben, professor of social biology at the London School of Economics, announced that it was "now legitimate to entertain the possibility that the human chromosomes can be mapped."[39] He and others regarded the heavily mathematical approach of mapping human genes through linkage studies as a way of turning genetics into an exact science and distancing it from speculative eugenic generalizations. At the same time, linkage mapping opened the possibility of identifying carriers of deleterious genes, a problem that had long vexed eugenicists.[40]

Although progress remained slow, the mapping of human genes remained an abiding interest for a dedicated group of human geneticists.[41] J. B. S. Haldane, evolutionary biologist, statistician, and "*the* moving scientific spirit" in genetic research at the time, most energetically promoted the project.[42] Delivering the Croonian Lecture at the Royal Society in 1946, he suggested that the "final aim [of human genetics], perhaps asymptotic, should be the enumeration and location of all the genes found in normal human beings."[43] On the other side of the Atlantic, McKusick, studying the Amish people, collected large pedigrees, especially including information on the distribution of hereditary diseases, which he submitted to elaborate statistical analysis in order to extract linkage information. In the late 1950s, he started using IBM computers at the Glen L. Martin Company, an aerospace firm (later known as

Lockheed Martin) with headquarters in the city, for the otherwise in-
tractable calculations. Eventually Johns Hopkins Hospital acquired its
own IBM computer. The programs for the linkage analysis were written by Jane Schulze at Johns Hopkins, in collaboration with James Renwick, who shuttled to and from London. McKusick also deployed the
computers to store and handle the information on genetic diseases he
was amassing for his *Mendelian Inheritance in Man*.[44] Despite his exceptional contributions to human gene mapping, McKusick once remarked,
"I am not certain why, in the late 1950s, I became enthralled with mapping genes on human chromosomes."[45]

Meanwhile the new work on human chromosomes provided hope that
genes could eventually be located on chromosomes through cytogenetic
techniques. The association of chronic myeloid leukemia with an unusually small chromosome (later identified as chromosome 22 that had
undergone a reciprocal translocation with chromosome 9) in 1960 can
be viewed as a first step in this direction, although it remained unclear
which genes were involved in causing the cell transformation. In 1968 the
linkage study between a chromosomal mutation (an uncoiled region of
chromosome 1) and the Duffy blood group, already known to be linked
to a congenital form of cataract, led to the first assignment of a human
gene to an autosomal chromosome. The feat was achieved in McKusick's
group.[46] By that time, sixty-eight human genes had been assigned to
the X chromosome through genetic linkage studies. Around the same
time, the development of two new cytogenetic techniques—chromosome
banding and a technique based on the construction of mouse-human cell
hybrids—opened the way for the mapping of human genes to proceed at
a much faster pace. Although banding techniques have been mentioned
before in various connections, their importance for mapping genes war-
rants a more detailed introduction here.

"Giemsa Magic" and Cell Fusion

Experimenting with DNA damaging (alkylating) drugs to study gene
activity during development, Lore Zech, working at Torbjörn Caspersson's laboratory at the Karolinska Institute in Stockholm, observed that
the antimalarial drug quinacrine mustard produced a fluorescent banding pattern along plant and other chromosomes. The pattern was repro-
ducible and characteristic for each chromosome. Zech and Caspersson

published a paper describing the banding pattern for all twenty-four human chromosomes, showing that the technique could be used to unam-
biguously identify chromosomes that so far had been difficult to distinguish.[47] Quickly, other groups achieved similar banding patterns using
Giemsa stain, a mixture of methylene blue and other dyes. The mixture,
named after the turn-of-the-century German chemist David Giemsa,
was originally devised to stain malaria parasites in blood preparations,
but it also became widely used for staining chromosomes. The differential staining showed up when the preparations were previously treated
with denaturing agents like heat, alkali, or detergents. Later, pretreatment with trypsin was found to be effective. Q-bands (produced by quinacrine) and G-bands (produced by Giemsa staining) were shown to
match each other closely. However, Giemsa staining had the advantage
that it was more stable than quinacrine fluorescence, which faded fast,
and that the bands could be seen with a standard light microscope rather
than requiring a more costly ultraviolet microscope. It quickly became
the standard technique for banding, although other mostly fluorescent
staining techniques soon joined the tool kit of cytogeneticists (fig. 5.3).[48]
With a nod toward molecular biology, darkly stained or brightly fluorescent regions in G-banding were interpreted as indicating regions of the
chromosomes with high adenine-thymine content, while less intensely
stained regions contained higher cytosine-guanine content. The four
bases were associated with the four nucleotides and their characteristic
pairing that made up DNA.

Besides facilitating the identification of chromosomes, the "Giemsa
magic" provided a number of landmarks along the chromosomes.[49] This
allowed researchers to detect and physically locate minor deletions,
translocations, and inversions that had gone previously unnoticed. The
new technique also fueled the study of chromosome organization, especially regarding the structure and function of the centromere region and
the distal ends of chromosomes that often showed the most characteristic
banding patterns. As one participant put it, "conventional cytogenetics"
became "obsolete almost overnight."[50] A standardization conference,
convened in Paris in 1971, provided elaborate guidelines for describing the newly observed chromosomal regions in an exact and reproduc-
ible way. It also provided standard diagrammatic representations of each
chromosome to facilitate comparisons (fig. 5.4).[51] The standard banding
chart offered human geneticists what drosophilists had all along: a physical map of the chromosomes on which genes could be located. In com-

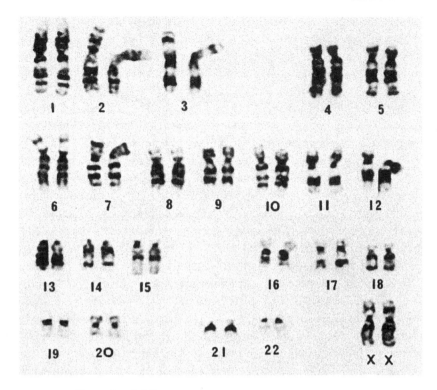

FIGURE 5.3. Giemsa-banded chromosomes.
Source: Hsu, *Human and Mammalian Cytogenetics*, 127, fig. 19.2. Reproduced with permission of Springer Nature.

bination with somatic cell hybridization, it provided a powerful tool for gene mapping.

Somatic cell hybridization relied on cell fusion, a technique that had been attracting wide interest since the 1960s.[52] Cancer cells grew particularly well in culture, and cancer geneticists at the Pasteur Institute in Paris tried to fuse malignant with nonmalignant mouse cells in an attempt to study DNA transformation and prove the chromosome theory of cancer. The researchers observed that fused cells grown in culture progressively lost chromosomes. Other researchers experimented with human-mouse hybrids (fig. 5.5). Using different stains for human and mouse chromosomes, they established that in such hybrids, human (rather than mouse) chromosomes were gradually lost. Chromosome banding made it possible to establish precisely which chromosomes or chromosome fragments were lost. Correlating these microscopic observations with the presence

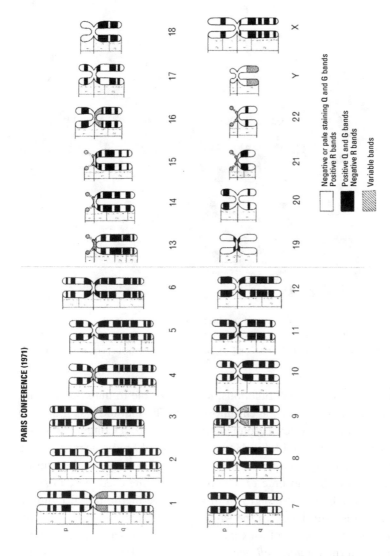

FIGURE 5.4. Diagrammatic representation of banded chromosomes.

Source: Hamerton, Jacobs, and Klinger, "Paris Conference," 334–35, fig. 5. Reproduced with permission of Karger.

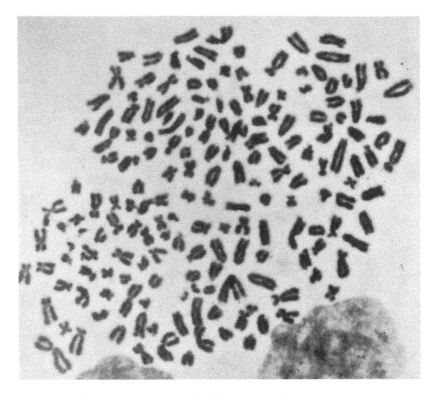

FIGURE 5.5. Chromosomes in a hybrid produced by the fusion of two human and one mouse tumor cells. The fused cell contains human and mouse chromosomes, 181 overall.

Source: Harris et al. "Mitosis in Hybrid Cells," 607, fig. 3. Reproduced with permission of Springer Nature.

or absence of specific cell functions, it was possible to map the relevant genes on specific sites of the chromosomes. The method was first used successfully in 1971, when a group of researchers from Columbia University, Yale University, and Johns Hopkins School of Medicine managed to map the gene for thymidine kinase, an enzyme involved in the synthesis of DNA, on chromosome 17. The mouse cell line used lacked the enzyme, and the hybrids were grown in a selective medium that required the enzyme for survival.[53] In the same year, Ruddle, who had been introduced to the concept of asexual genetic analysis—for which he later coined the term *somatic cell genetics*—by Guido Pontecorvo during a postdoctoral year at Glasgow University, published the chromosome assignment for the genes of seventeen human enzymes in a single article.[54] A "torrent of papers" reporting many more gene assignments using the same method followed in the next few years.[55] Theoretically, the locus of any biochemical product in the cell that could be (electrophoretically) distinguished from its mouse counterpart could be mapped in this way. Though cumbersome, the method circumvented sexual reproduction and genetic-crossing experiments and did away with the complex statistics of linkage studies. For all these reasons, it soon became the method of choice for human gene mapping. With the wealth of clinical data available, progress on the human gene map soon outstripped the mapping projects of traditional model organisms.

A Complete Map of Human Chromosomes

With the number of gene assignments growing rapidly, Ruddle convened the first International Workshop on Human Gene Mapping in New Haven in 1973. The workshop gathered sixty-eight participants. Many came from nearby North American institutions—indicating a shift of activity from Europe to the United States—but representatives also came from Canada, the United Kingdom, the Netherlands, and France. Apparently the number of participants was considered too high, as those present agreed that future participation would be by invitation only. Moreover, it was stipulated that not more than two people from any center were allowed to attend future meetings. Only the host institution could have more delegates. The overall number of participants was not to exceed forty, with fifty as the "absolute limit."[56] Funding for the first and all following meetings came from the National Foundation, which by this time had redefined its mission as the study and prevention of birth defects as reflected in the name change, in 1976, to March of Dimes Birth Defects Foundation.

In the introduction of the first meeting report, published in the *Birth Defects Original Article Series*, funded by the National Foundation, and reprinted in *Cytogenetics and Cell Genetics*, the editors underlined the promises of somatic cell genetics for the compilation of a human gene map. In conjunction with family studies, "the acquisition of new data has been rapid—and promises to become explosive." From the beginning the aim was to make the workshops an annual event (later a biennial one) to "evaluate progress and chart new directions." A distinctive feature of the workshops was that the meetings were used for actual work on the map. Committees assigned to review particular areas of the genome confirmed valid gene assignments and flagged those that needed further confirmation. Participants also agreed on a standard terminology. In addition, there were discussions on new concepts and methodologies. Overall the meetings were designed to contribute to "the more orderly advance of human gene mapping."[57] From the beginning the aim was "to map all the genes," a goal to be achieved by the year 2000.[58]

Subsequent meetings took place in Rotterdam (1974), Baltimore (1975), Winnipeg (1977), Edinburgh (1979), Oslo (1981), Los Angeles (1983), Helsinki (1985), Paris (1987), New Haven (1988 and 1989), Oxford (1990), and London (1991). The meeting reports document the rising number of gene assignments and the growing scale of the overall enterprise. The number of assigned genes rose from just over 200 at the first meeting to more than 2,300 genes at the last meeting, when the human genome sequencing project started to take off, with the increase of assignments doubling every year.[59] Special committees were responsible for reviewing the gene assignments to specific chromosomes. Initially, only the X chromosome and chromosome 1 had their own committee, while another committee was responsible for "all autosomes other than chromosome 1."[60] With the growing number of genes attached to each chromosome, the number of chromosome committees expanded.

Along with the assigned loci on the chromosome map, the reports listed the method used for mapping. Initially, somatic cell hybridization and family linkage studies provided the bulk of new assignments. However, already at the first meeting, participants discussed the potential contribution of in situ DNA-RNA or DNA-DNA hybridization for gene mapping. Just a few years earlier, Mary-Lou Pardue and Joseph Gall from Yale University had reported the hybridization of a radioactively marked DNA probe to a cytogenetic preparation of the toad *Xenopus*.[61] The scientists convened at Yale agreed that if the method became "practical," it could develop into a "powerful tool" for the localization of genes on chromosomes.[62] Every method that promised to contribute to the cartography of human genes was welcomed.

Pardue and Gall had prepared the test DNA for their experiment by isolating it from a tissue culture. The development of methods for cutting, pasting, and cloning DNA gave in situ hybridization new scope. The Oslo meeting in 1981 "left very little doubt that recombinant DNA techniques will have an unprecedented impact on human gene mapping."[63] A new committee was created to review progress in that area. At the same meeting McKusick, who had taken on a leading role in compiling the map, in his intervention stressed the advantage of using multiple methods in a cooperative way.[64] In particular, classical linkage studies were attracting new attention as part of the "recombinant DNA revolution" as linkage groups could now be mapped to concrete locations on the chromosomes and molecular markers provided additional factors for linkage analysis.[65] Similarly, recombinant DNA techniques could be used in combination with or parallel to somatic cell genetic techniques.

Young molecular biologists who were at the meeting took a more confrontational stance. Edwin Southern from Edinburgh, the inventor of "Southern blots," a key technique for separating and selecting specific DNA fragments through hybridization, regarded the introduction of molecular technologies as a "break with the traditional methods of genetics." He also pointed to the "gap" in terms of scale between the lengths and positions of DNA fragments on the "molecular map" and "the chromosomal map," with its banding pattern.[66]

At the same meeting, a new mapping method was presented to address that problem. The idea, first launched by David Botstein and his team at the Massachusetts Institute of Technology, was to use arbitrary reference points, like DNA markers, rather than a band or a gene, for linkage and DNA mapping, and in this way to bridge the gap between map positions based on cytogenetics and mapping at the DNA level.[67] The provision of new markers along the chromosomes was seen as perhaps the most important contribution of molecular biology to mapping. The markers were especially useful for map-based gene discovery. Using this method, the gene for Huntington's disease, a dominant hereditary condition, was located in the early 1980s.[68] For clinicians like McKusick, the aim of being able to detect genes associated with diseases before clinical symptoms were apparent was integral to the mapping effort. It gave clinical significance to the map as well as the catalog of human diseases he was compiling. By the same token, the advancement of medical genetics was predicated on the advancement of mapping.

Meanwhile, the increasing volume of data made the computerization of the map a pressing issue. With the map growing from a one-page list of gene assignments to a densely written text of over one hundred pages, the text format was increasingly viewed as "unsatisfying" for updating or searching the document. For several years, Ruddle, the initiator of the mapping conferences, received funding, first from the NIH and later from the Howard Hughes Medical Institute, to digitize the mapping information he was collating and build up a database. If, at the first meeting in 1973, "informatics was a science of pencil and paper notation," fifteen years later, mappers—convened again under Ruddle's chairmanship at Yale for an "interim meeting"—relied for their work of editing the map on "complex computer programs and fifty computer workstations interconnected with a main frame computer."[69] One year on, workshop participants could built on a "truly interactive, on-line database" that they could "modify directly as well as browse."[70] The database was also accessible remotely by telephone, by Telenet line, and through personal computer terminals across the United States and, with some success, in Canada and Europe.[71] In the late 1980s, the Human Gene Mapping Library, housed at Yale, was the world's largest human gene mapping database, with a staff of about nine people working on it full-time.[72] At about the same time, the database of genetic diseases that McKusick had been building up since the mid-1960s also became available online. The database, known as Online Mendelian Inheritance in Man, or OMIM, came equipped with enhanced search capabilities. Pointers provided links between the two databases.

However, the meeting in New Haven in 1989 marked an even more radical transition point for chromosome mappers. The growing importance of molecular techniques for mapping attracted new participants. More important still, scientists had come to agree that the "complete mapping and sequencing of the human genome," in a period comparable to that between the first and the tenth Human Gene Mapping Workshop, or about fifteen years, was "both feasible and desirable."[73] The plan was not without its critics, but the US government had committed itself to the project. The expanded effort required a new organization and a full-time permanent office that would be supported by the Human Genome Organization (HUGO). The Human Gene Mapping Workshops would be replaced by single chromosome workshops and an annual chromosome coordinating meeting, associated with a Human Genome Mapping Workshop. A new database, the Genome Data Base, housed at Johns Hopkins, would serve the new project.

The transition to the new organization under HUGO was completed at the next meeting in London two years later. At the preparatory meeting in Oxford the year before and again in London, efforts were centered on the launch of the database. Genome Data Base allowed for both physical and cytogenetic mapping information to be entered. It could not only be accessed but also edited remotely—a major technical improvement over the earlier model. A separate program was able to prepare "publication-quality idiograms" that accurately reflected the relevant gene-mapping information. A further advantage was that the database could be linked to other databases, in particular to OMIM and the mouse-mapping database. It is noteworthy that several software companies mounted exhibitions or otherwise advertised their services and products at the meeting. This was the first time such activity was mentioned in the reports, possibly the beginning of a new trend.[74]

At least initially, Ruddle, McKusick, and other cytogeneticists who had been active in convening the gene-mapping conferences, welcomed the new plans that they saw as a continuation of their own efforts.[75] Ruddle especially was centrally involved in early discussions on the plan to map and sequence the human genome. He acted as an effective chairperson of the Genome Sequencing Workshop in Santa Fe, New Mexico, in 1986, a key planning meeting convened by the Department of Energy and the Life Sciences Division of Los Alamos National Laboratory. He also participated in related meetings organized by the NIH and the Howard Hughes Medical Institute, and, with McKusick, he was part of the National Research Council Committee on Mapping and Sequencing the Human Genome, which recommended that Congress allocate $1 billion in new funds to the mapping project.[76] In 1987, together with McKusick, he founded the journal *Genomics*, dedicated to the publication of work on the mapping and sequencing of (human and other) genomes. "Genomics" was presented as a new discipline, but the editors did not fail to highlight the connection with the earlier gene-mapping effort they had actively pursued since the early 1970s. The "nucleotide sequence" was presented as "the ultimate map" and a useful step toward gene mapping, a goal to which they remained committed.[77] With the sequencing project gaining steam, Ruddle, together with a small handful of other candidates, was informally considered as possible director of the new project, while McKusick became the founding president of HUGO, the coordinating agency of the international effort to sequence the human genome.[78] Ruddle supported the construction of a physical map of the whole genome and the sequencing of clinically relevant regions as well as comparative sequencing. He also argued for a distributed structure with centralized funding and management. Although some of these ideas became an integral part of the publicly funded Human Genome Project, by the late 1980s, Ruddle's influence started to wane and card-carrying molecular biologists and directors of designated sequencing centers gained increasing influence. The decisive blow was the move of the mapping database from Yale. Following critiques regarding the way the database was constructed—on a hierarchical rather than a relational model, as was becoming the norm—and its inadequacy for the tasks ahead, funding by the Howard Hughes Medical Institute stopped abruptly.[79] Efforts to rebuild and continue the database at Johns Hopkins were short-lived, and the whole project "vanished" while GenBank at Los Alamos National Laboratory and other databases in Europe and Japan took over as information repositories for the Human Genome Project.[80]

Despite the diminishing influence of cytogeneticists and somatic cell geneticists, the continuities between the Human Gene Mapping Workshops and the later organization supporting the international Human Genome Project were evident—down to the use of the same abbreviation, HGM, for the workshops supporting both initiatives. The two projects also shared the international collaborative structure, the data-sharing arrangements, and the noncommercial aspect. In the view of the "gene mappers," the whole genome project depended—technically and politically—on the previous chromosome-mapping efforts. As Ruddle put it in a later interview, "I don't think Congress or anyone would have accepted [the genome project] without the realization that many genes had already been mapped and that there was progress."[81]

Nevertheless, invited to deliver a keynote lecture at the 1998 human genome meeting, McKusick complained rather bitterly that the promoters of the project were "not familiar with what had gone on in the field of gene mapping." Like others, he remained critical that the Human Genome Project, as proposed by a handful of leading molecular biologists, was purely based on sequencing, without any reference to gene mapping, which had been the focus of activity before. To underline his point, McKusick declared that, while James Watson wanted the Human Genome Project to be finished by April 2003, or fifty years after he and Crick first proposed the double helical model of DNA, he personally would be satisfied if the project were not completed until 2006, or exactly fifty years after the correct chromosome number was established.[82]

As it turned out, the available genetic map proved essential to constructing the physical map of the human genome and aligning the many DNA fragments that composed it, an essential step in producing the full sequence. Sequence annotation, including gene assignments—with all the complications attached to defining what a gene is—became a central preoccupation in the post-genomic era.[83] In this endeavor, the chromosomal maps created by cytogeneticists continued to serve as visual reference tools for genomic researchers and medical geneticists alike.[84]

Whose Turn?

Discussions between molecular biologists and gene mappers at the workshops reflected more general tensions that were playing out in various research groups. Genetic engineering, together with DNA sequencing techniques, offered molecular biologists a new set of tools that made it possible to tackle questions of human heredity in the test tube. The invention, in the mid-1980s, of the polymerase chain reaction (PCR), a powerful biochemical technique to amplify DNA probes, further expanded the technical possibilities. Funding opportunities and hopes for medical returns further encouraged molecular biologists to move from their entrenched work on model organisms to work on humans.[85]

A new generation of medical students also became increasingly interested in genetics as a medical specialty. Gene mapping especially seemed to hold promise for diagnostic tests and eventually therapy. The promises of the new DNA techniques attracted young people to the field. Yet there were obstacles to choosing such a career. Medical students were discouraged from following that route. They were told nobody needed clinical geneticists and that it was "a bad choice."[86] Only a few places offered training possibilities. McKusick's clinic in Baltimore was one of the attractive options. It received a constant flow of researchers from the United Kingdom and elsewhere.[87] In contrast, the University of Cambridge in England, for instance, long resisted introducing human genetics in its teaching curriculum or supporting research in the field. The clinical genetics service in the area, too, remained patchy well into the 1980s, when a new effort was made to build up medical genetics at the new clinical school and expand the service.[88]

Nevertheless, what is often described as cytogenetics "going molecular" could just as well be described as the turn of molecular biologists (and clinicians) to human and medical genetics. Historians have pointed to the difficulties molecular biologists encountered in their quest to expand their investigations from viruses and bacteria to higher organisms.[89] Yet the often-contested turn of molecular biologists to human and medical genetics is generally not articulated or is made to appear as the next logical step, as in the move from the worm genome to the human genome sequencing project.[90] And yet the study of the human genome was a field long occupied by human chromosome researchers. In the 1980s, the pressure on molecular biologists to deliver on long-promised medical payoffs was mounting, and studying the human genome could be viewed as moving toward that goal.[91] Not surprisingly, then, the human genome sequencing project was presented as a decisive step toward conquering cancer and other genetic diseases, even though arguably it was above all a large technological project, meant to speed up sequencing rather than putting clinical concerns first.[92]

How did cytogeneticists react to the new incursions into their field? As noted at the beginning of this chapter, some cytogeneticists embraced the new molecular technologies. Among them was Malcolm Ferguson-Smith. A pathologist by training, he became fascinated with human chromosome research in the late 1950s. After a stint at McKusick's clinic in Baltimore, he returned to Glasgow to build up medical genetics at the university there.[93] Recognizing the potential of recombinant DNA technologies for mapping genes, he collaborated with other scientists, hired young people trained in the techniques, and learned them himself. A trainee in his department remembered: "When I arrived, Ferguson-Smith was running a blood grouping laboratory, a protein polymorphisms laboratory, a cytogenetic laboratory and a clinical service. Around 1983, he shut down the former two and opened a DNA lab."[94] Although Ferguson-Smith embraced molecular technologies, he insisted that cytogenetics was "at the heart of modern genetics" and that this would continue to be so. As he saw it, "molecular biologists and others who ignore cytogenetics do so at their peril as, without it, they are likely to have an incomplete understanding of the fundamentals of genetics."[95]

The need to "go molecular" was generally recognized, but the transition often produced tensions.[96] For instance, at the MRC Clinical and Population Cytogenetics Unit at Western General Hospital in Edinburgh, a new Molecular Genetics Section, headed by Southern, was instituted in 1980.[97] Scientists in the section used molecular techniques to study the structure of chromosome bands; they used DNA hybridization techniques to locate specific sequences on the human genome, and in collaboration with the MRC Mammalian Genome Unit at Edinburgh, they worked on the construction of a restriction site or molecular (rather than a chromosomal) map of the human genome.[98] They also worked on the identification of disease markers. In addition, scientists in the Molecular Genetics Section were encouraged to interact with the other groups in the unit and help them "go molecular."[99] The pressure to move more quickly in this direction mounted with subsequent site visits by MRC review committees. Cytogeneticists pursuing long-established studies on a wide variety of clinical populations felt increasingly sidelined and "made to feel like people from a bygone era."[100] For some it was indeed difficult to understand the "new language" of molecular biologists and to see how they could integrate the new techniques into their work. They nevertheless saw value in continuing their studies on sometimes unique data sets or patient groups. Yet when they applied for postdoctoral researchers trained in the new techniques to collaborate on gene assignments in their family data, the projects were more likely to be simply passed on to newly hired molecular biologists, who often preferred working on animal models rather than human data sets. This inevitably led to resentment. Cytogeneticists would have liked to see "lower order" techniques combined with "higher order" knowledge and expertise they had accumulated over decades of painstaking work. Instead, cytogeneticists were encouraged to leave or take early retirement: "One by one the microscopes were put down in the stores."[101] The registry with the patient data, once the backbone of the work in the unit, first moved into a smaller room and eventually (possibly around 1998) was closed down.

Nevertheless, some molecular biologists did recognize how important the work on chromosomes was for molecular studies. Nicholas Hastie, who joined the unit as a young group leader in the molecular biology section in the early 1980s and later became director of the newly renamed Human Genetics Unit, acknowledged: "The linkage maps, the physical map and genomics would never have happened without the prior work on chromosomes. Equally, much of the work on deletions and translocations that were used to identify diseases started from observations of the chromosomes. The cases and the data were taken directly from the registry started by Court Brown who thought ahead of his time. . . . Later work relied on the registry in respect to both the data and conceptually."[102] The last comments related to the fact that the genetic, medical, and family data collected in the registry helped to identify various genes, including, for instance, a series of genes responsible for the development of the eye (the *PAX6* gene and other members of the *PAX* gene family), for the susceptibility to bowel cancer, and *DISC1*, a gene responsible for brain development and brain synapses.[103]

For some of the young molecular biologists sent out to find a new workspace among the cytogeneticists, the view through the microscope came as a revelation and set the course of their future research. Wendy Bickmore joined the MRC Clinical and Population Cytogenetics Unit as part of the new intake of molecular biologists. She had been working with DNA maps and gel electrophoresis trying to find variations in DNA fragments and locating genes. When invited by her new colleagues to look at a chromosome preparation through the microscope, she was captivated. She found that observation through the microscope provided an "immediacy" and "a feel for processes in the cell" that no other technique offered. She became interested in studying the molecular signature of banding patterns as well as the spatial organization of chromosomes in the nucleus and its connection to gene expression in development and disease. Both research areas were directly inspired by looking at cells through the microscope, yet, especially the spatial arrangement of chromosomes in the cell, had long been neglected in research.[104] The studies became part of a more general renewed interest in the structure and regulatory functions of chromosomes and their role in health and disease.[105]

A cartoon circulating among cytogeneticists shows a molecular biologist declaring from a lectern in 1977, "As a renowned scientist I can safely say that the field of cytogenetics is dead." The same pronouncement is repeated in 1987 in front of a display of Southern and other (imaginary) blots, and again in 1997. This time the clearly aging molecular biologist, speaking at the "Molecular Human Molecular Association of Molecular Genetics," stands in front of a display of polymerase chain reaction (PCR), restriction fragment length polymorphism (RFLP), and other diagrams. In the final vignette for 2007, the *Journal of Applied Cytogenetics* announces the death of the renowned molecular scientist (fig. 5.6). As this comic strip, penned by Richard Sherman, a cytotechnician with a sense of humor, suggests and other cytogeneticists confirmed, cytogenetics not only has survived but also is celebrating something of a comeback in the post-genomic era, most notably in cancer diagnostics and the highly lucrative field of cancer drug development.[106] The DNA changes in cancer chromosomes are too complicated and unstable to be studied with standard molecular techniques. This leads to the seemingly ironic consequence that, in clinical practice today, molecular geneticists perform routine genetic diagnoses while cytogeneticists are entrusted with the complicated analysis of somatic cancer cells. Similarly, drug companies ask to see the targets of cancer drugs that are

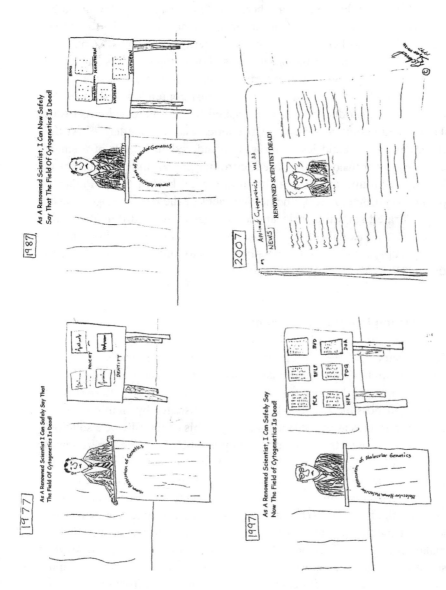

FIGURE 5.6. Cartoon by Richard Sherman, originally published in *Applied Cytogenetics* 23, no. 1 (1997): 19. Source: Courtesy of Sibel Kantarci (UCLA). Reproduced with kind permission of Richard Sherman.

being developed, which again requires cytogenetic techniques. FISH techniques, the fluorescent marking techniques mentioned at the beginning of the chapter, are among the tools cytogeneticists employ for this work. Interestingly, the key case here was Gleevec, a drug developed by Novartis for the treatment of chronic myeloid leukemia in the 1990s that was celebrated as heralding a new era of therapeutic cancer agents produced using molecular biological tools. For cytogeneticists, the story marked an altogether different turning point. "Cytogenetics seemed on the way out," a still-practicing cytogeneticist contended, "when in the late 1980s the producers of Gleevec asked to see where the target was. This forced people to see that cytogenetics was still needed."[107] At the time, chronic myeloid leukemia was considered the result of a single chromosome translocation between chromosome 22 and chromosome 9, leading to an unusually short chromosome 22, known as the Philadelphia chromosome. During the development of the drug, cytogeneticists demonstrated several translocations that gave rise to different protein products. This explained the different effect of Gleevec on different patient groups, and it again became important to understand resistance to the drug.[108]

The Triumph of the Microscope

Disciplinary disputes like the ones documented here between chromosome researchers and molecular biologists are widespread in the sciences. Intellectual claims, funding streams, and career patterns are at stake. Is it possible to move beyond these disputes by simply focusing on chromosomes as the object of inquiry? This can hardly be a solution, as "epistemic objects" do not exist independently from the experimental systems in which they are embedded.[109] Chromosomes therefore always point us back to the technical conditions of their respective representations and to the same disciplinary disputes. Thus, we see Crick bluntly defining the metaphase chromosomes, the classical object of cytogeneticists, as the "dullest form of chromosome: an inert package needed to make orderly mitosis possible," while recognizing only a molecular description of genetic information.[110] Cytogeneticists rebutted, subsuming much of the twentieth-century history of genetics, from orcein chromosome staining to DNA sequencing, under the rubric of "chromosome research" and, thus, "cytogenetics." Antonio Lima-de-Faria, professor emeritus of molecular cytogenetics at Lund University and author of the encompassing volume on cytogenetics, followed suit with another book in which the chromosomes tell their own convoluted story.[111] Cytogeneticists interpreted the book as making their case. Wrote Ferguson-Smith: "I firmly believe that molecular biologists need an understanding of the nature and behaviour of chromosomes if they are to fully advance in their discipline. I believe that your book will give them the necessary direction."[112]

However, the story of chromosome structure and mapping told in this chapter also points to the epistemic and historical commonalities between the two contending approaches that developed at the same time. Chromosome researchers and molecular biologists shared an interest in the material basis and the mechanism of heredity as well as in the nature of mutations, a vital concern in the nuclear age. Together they established genetics as a key issue in biology and biomedical research. Both the chromosomes and the letters composing the DNA code are presented as the "alphabet" of life, the language in which the book of nature is written and that in the end defines "us," as a species and individually.[113] Thus, when the artist Suzanne Anker and the sociologist Dorothy Nelkin discussed the scriptural metaphors of genetic information and their artistic renderings, they could draw indiscriminately on DNA and chromosomes as examples.[114] Similarly, the compact disc that DNA sequencing pioneer and enthusiast Walter Gilbert pulled out of his pocket to demonstrate how people will carry their personal genetic information with them is matched by the personal karyotypes exchanged by chromosome researchers as "self-portraits," visiting and greeting cards thirty years earlier.[115]

Furthermore, chromosome researchers and molecular biologists shared the assumption that the study of structure would explain function. Pattern recognition, mapping, and the harnessing of computers as both memory and calculating machines played a role in both endeavors.[116] Indeed, imaging techniques like X-ray diffraction and electron microscopy, including recent advances in cryo-electron microscopy, have been at the center of much research in molecular biology. Similarly, both cytogeneticists and molecular biologists seized first on radioactive markers and later on fluorescent tagging to visualize structures and functions. Intriguingly, cytogenetic observations with radioactive labeling provided the first evidence for a semiconservative mechanism for the replication of the genetic material, a finding usually attributed to the experiments, using molecular techniques combined with radio-active tracers, of Matthew Meselson and Frank Stahl.[117] On their part, molecular life scientists compared the significance of radioactive tracer techniques to the introduction of microscopes.[118] Fluorescent labeling perhaps more than any other technique has confounded the usual distinctions between the two endeavors. If FISH and comparative genomic hybridization techniques, both based on fluorescent markers, allowed chromosome researchers to "go molecular," fluorescent tagging in conjunction with vastly refined and digitally supported microscopic techniques have been changing the working practices and research objects of molecular biologists.

Since the 1950s molecular biologists had relied on the fractionation of cells, radioactive marking, photographic techniques, and model building to visualize, track, and localize molecular processes. The possibility of specifically tagging molecules with fluorescence-marked antibodies—and, even more, the introduction of fluorescent proteins produced by the cell itself, like the green fluorescent protein first described by Martin Lee Chalfie from Columbia as a marker for gene expression—in combination with the development of a new generation of microscopes that greatly enhanced the contrast and resolution of fluorescent imaging provided a powerful new approach to study molecular processes.[119]

Important effects were achieved using laser as a light source. In the confocal laser-scanning microscope widely used in biology, light is focused on the specimen in one spot at a time and recorded point by point. A pinhole in front of the detector allows only light from the focus point to be recorded. This reduces out-of-focus glare and produces much sharper images than traditional wide-field fluorescence microscopes. The object can be optically sectioned and scanned layer by layer at different focus positions. Computer software reconstructs the images in two and three dimensions. Colors are added to the black-and-white fluorescent images via computer algorithms, which allows for the creation of the colorful pictures that now fill the pages of molecular biology journals.

The principle of the confocal microscope was already understood in the 1950s, but it was only in the mid-1980s, with growing demand for fluorescence microscopes for biological research and considerable developments in computer hardware and software, that a prototype adapted to the study of biological specimens was developed and commercialized.[120] Although the electron microscope reaches higher resolution, the confocal microscope has the advantage that it allows for live-cell imaging. More recently, wide-field fluorescence microscopes using computer-enhanced imaging processing offer an alternative to the confocal microscope. Meanwhile, super-resolving fluorescence microscopes that have broken the theoretical limit of image resolution make it possible to visualize molecular processes like DNA transcription as they happen.[121] By being able to visualize genetic processes in single cells rather than studying average processes in a population of cells, molecular biologists now do what chromosome researchers have always been doing: they look at the genome of one cell at a time.[122]

As chromosome researchers all along, molecular biologists now also study microscope images, while fractionation techniques, analytical centrifuges, and scintillation counters that—together with X-ray diffraction and electron microscopy—dominated molecular research in the 1950s to 1980, have moved out of favor.[123] Even the Illumina sequencer, widely used for genomic sequencing, is based on the principle of a fluorescence microscope. The instrument tracks the addition of labeled nucleotides as the DNA chain that is being sequenced is copied in multiple parallel processes. More generally, the new visualization techniques inspire a new "fluorescent aesthetic" of biology that spills over in the public visual culture of science.[124]

In the historiography of the postwar life sciences and biomedicine, molecular biology has taken the limelight. The argument presented here is that the microscopic techniques of cytogeneticists and the molecular techniques both contributed to the study of the human genome. Or, to paraphrase Crick, looking at pictures and reading texts were both essential.

Epilogue

In the late 1950s chromosome analysis emerged as a powerful tool to study heredity in humans. Unlike any other technique, the improved chromosome preparations offered a direct glimpse of the complete genome of an individual, opening up endless possibilities of observation and intervention. To the scientists who, for the first time, were able to look at their own chromosomes, this felt like "walking on the moon."[1] And just as space exploration opened up a new view of planet Earth, the study of human chromosomes promised new insights into how heredity works and how genes make us "who we are."[2] Chromosome analysis quickly supplanted other genetic techniques, notably pedigrees and blood groups, as the key technology for the study of human heredity and provided much of the impetus for transforming human genetics from a "backwater" to a dynamic "research frontier."[3] As radiobiologists, pathologists, clinicians, cancer researchers, patients and patient advocacy groups, anthropologists, lawyers, Olympic committees, athletes, policy makers, and activists embraced and contested the technologies and the genetic explanations that came with them, the meanings of chromosomes and of human heredity expanded and changed.

Chromosome analysis provided insights into the structure, duplication, and transmission of the genetic material. Spurred by the observation of chromosome anomalies such as trisomies, scientists probed the mechanisms by which chromosomes are distributed into germ and daughter cells and how this process can go wrong. Observations of the distal ends of chromosomes provided insights into satellite formation and the loss of chromosome material through aging. Centromeres, too, were studied in detail. Most notably, chromosome techniques, in conjunction with somatic cell fusion techniques, were employed to map human genes, circumventing the need for cycles of sexual reproduction that were difficult to observe in humans. At the same time, chromosome analysis could be used to track heritable changes running in families as well as changes that occurred during development or later in life through various exposures or disease events—in all cells of the body. This entailed a considerable expansion of the reach of the techniques of human genetics, now covering horizontal (as, for instance, in cancer cells) as well as vertical transmission (through the generations).

Mutations of all kinds could be made directly visible. On the basis of this principle, chromosomes were turned into biological dosimeters for atomic radiation and other mutagenic workplace or environmental toxins at a time of rising public concerns about atomic weapons testing and environmental pollution. Chromosome observations led to the establishment of a new class of diseases—the chromosome diseases, first so named by Lejeune and his colleagues in Paris in their publication on trisomy 21 in 1959. Chromosome diseases could affect mental abilities and behavior, morphological traits, and sexual development. Specific forms of cancer, too, were reclassified as genetic diseases. Chromosome diseases could be diagnosed in adulthood or at birth as well as prenatally, shifting the diagnosis from clinical symptoms to interpretable chromosome images and suggesting a new set of interventions made possible through changes in reproductive rights legislation. Yet chromosome analysis also made an impact outside the clinic. It was employed for gender identification in the competitive sports context, and in the courts to argue for mitigating circumstances in murder trials. Traveling relatively easily, the techniques, often in tandem with other approaches, were employed in surveys of defined human groups and worldwide population studies in elusive attempts to trace common evolutionary mechanisms as well as biological markers that distinguish specific groups. Navigating the complex and fraught terrain of human heredity, chromosome researchers, peering down the microscope while employing ever-refined techniques for visualizing differences, established networks and infrastructures that could later be redeployed for genomic studies, once molecular scientists turned their attention from the study of simple organisms to the study of human heredity. Such transitions, however, were always contested and had to contend with changing criteria of value and justification.

Researchers celebrated that chromosome analysis provided a firm basis for the study of human heredity. Yet the expansion of genetic technologies around chromosomes and the increasing reach of genetic explanations were resisted at every step. The use of chromosome analysis for prenatal diagnosis was vehemently denounced by some of the very researchers who contributed to developing the techniques that made it possible. Despite its broad adoption, prenatal diagnosis and the indications for abortion it provided remained controversial for ethical and religious reasons and for the principle of defining the worth of human life on the basis of biological differences. This raised concerns that eugenics would return through the "backdoor."[4] The prospective studies of children with defined chromosome anomalies, though well intentioned, were denounced as "bad science," biased and dangerous. In the heady days of the 1960s, the same critics exposed and protested the attempts to search for biological explanations of social problems and to take this as a basis for any kind of social intervention. Similarly, chromosome testing for gender identification was widely regarded as inappropriate. Applied to only female athletes, it remained controversial for the very results it produced as well as for the expectation that gender could be defined by a "simple" biological marker. Indigenous people, who were so widely bled in the large-scale screening projects of the postwar years, also increasingly protested these practices and claimed "genetic sovereignty."[5] Apart from questions of identity and sovereignty, commercial interests intervened in these debates. By the 1990s, the collecting of blood samples and genetic data as practiced in the preceding decades was widely regarded as unethical, leaving the status of entire collections in limbo.[6] The controversies highlight the many ramifications of chromosome research and the contested nature of human heredity in the long 1960s and well beyond.

Much of the excitement and fascination with chromosomes was based on the visual evidence the chromosome preparations provided. Researchers and technicians studying chromosomes under the microscope were as impressed by what they saw as other professionals, individuals, patients, and artists looking at chromosomes—in squash preparations or lined-up order, on photographs, projected in lecture halls, in medical reports, or in the media. Appropriately pointed out and explained, the specific characteristics concerning the number, shape, and staining of the chromosomes were there for all to see, even if their original observation required skill and expertise. Exposure to radiation or other mutagenic toxins including viruses resulted in an array of visible and quantifiable changes in the number and shape of chromosomes. Complex disease syndromes could be reduced to the visible loss or the addition of specific chromosomes or pieces of it. In more recent experiments or diagnostic tests, fluorescent probes mark the presence or absence of specific molecular sequences along the chromosomes on the computer screen. The visual representation of banded chromosomes continues to guide genomic assignments.[7]

Yet the reliance on visual evidence also marked the weakness of chromosome analysis—at least in the eyes of its detractors. Observations under the microscope, even if guided by theory and experiment, have often been dismissed as being "merely" descriptive.[8] There is an equally long tradition, especially in the physical sciences, of spurning images in favor of figures, mathematical formulations, and causal explanations.[9] When Crick admonished cytogeneticists to not just look at "pictures," he revived that view. Without any doubt, molecular biologists, too, relied on visual evidence in their work—or so it appears. It suffices here to recall the role that the X-ray diffraction image of DNA (now known as photograph 51) taken by Rosalind Franklin and her assistant Raymond Gosling played in the work that led to Watson and Crick's proposal of the double helical structure of DNA. The two researchers' (unacknowledged) use of the photograph has been widely condemned. Such a polemic would not have ensued had the picture not provided useful evidence. Indeed, to the trained crystallographer, the picture pointed to a helical conformation. Yet Crick was quick to note that Watson, the only one of the two to see the picture, did not have sufficient mathematical and crystallographic knowledge to fully grasp its meaning.[10] This points to an important difference in the way molecular biologists and cytogeneticists viewed images. For molecular biologists, what counted were the numbers that could be deduced from the images. The intensities of the diffraction spots were measured and the figures used for crystallographic calculations. Epistemically, cytogeneticists honed observational skills while crystallographers focused their attention on the amount of calculations required to move from one step of their analysis to the next. Molecular biologists appealed to images only when supported by extensive measurements and calculations. Although fascinated by the chromosome pictures and the meaning they conveyed, Penrose, trained in mathematical genetics and strongly committed to raising the scientific status of human genetics, also insisted on measuring chromosomes rather than just relying on visual evidence for their identification. Yet his approach met with resistance from other chromosome researchers who disputed the usefulness of measurements in the study of chromosomes.[11]

The appeal to numbers to justify visual evidence has not abated—evidence to the contrary notwithstanding. Publications in molecular biology today are awash with colorful illustrations, and new microscopes that enable researchers to visualize molecular structures and processes are creating a buzz in the field. Current experimental work that probes the three-dimensional molecular structure of chromosomes and its regulatory functions relies on the same imaging techniques. Yet the images these instruments produce are the product of much number crunching. In the new generation of microscopes, software algorithms that analyze and recompose the images from vast amounts of data are crucial for increasing the optical resolution of the instruments. Colors, too, are added to the black-and-white fluorescent images via computer algorithms. Furthermore, what counts is the quantitative information, the graphs and figures that can be pulled out from the imaging information. In many respects, microscopes become "quantitative tools."[12] A publication without figures and graphs would be difficult to publish. The colorful images have a rhetorical function and provide proof of the experimental procedure but are not essential for the communication of the results.[13] Bioinformatics, too, increasingly relies on images to visualize data and make them amenable to examination and manipulation. Yet despite the reliance on images, database research is considered theoretical work.[14]

The different meaning attached to visual evidence can explain the sidelining of human chromosome research in historical accounts of twentieth-century genetics, especially as these have been written predominantly from the vantage point of the triumphs of molecular biology. The closer inspection of the history reveals the assumptions that guide such interpretations and makes it possible to explain how, despite all the friction, the microscopic study of chromosomes and the analysis of DNA could develop and expand at the same time and share some common ground.[15] The observation of fixed individual chromosomes under the microscope, detached from their in vivo context, can be regarded as having prepared the ground for their description in molecular terms. The focus on structures and processes in the cell; the attention to questions of human and medical genetics; the expectation that many diseases, including cancer, have a genetic basis; the importance attributed to population studies and large data collection projects; together with some of the pushbacks and ethical debates that characterized the work of human chromosome researchers all along continue to play a central role in today's biomedicine. The study of chromosomes like that of DNA both participated in the triumphs and darker sides of the (past) "century of the gene."[16] In the post-genomic era, heredity has again become a more fluid concept.[17] As biology, helped by new visualization techniques, analyzes more integrated and dynamic processes in the cell and the wider milieu, the understanding of what makes us who we are is changing again. The genome is seen as multidimensional and adaptive, as reactive rather than as agent.[18] Yet in the current search for new understandings, the dynamic three-dimensional structure of chromatin fibers (before they are packed into chromosomes) and their role in regulating gene function, answering to environmental stimuli written down in epigenetic marks, once more promises to hold the key. Then and now, chromosomes stand at the crossroads of a cellular and a molecular understanding of life.

Acknowledgments

W hile researching and writing this book I have relied on the help and support of many people and many institutions. It is a pleasure to thank them here.

Work on the book started during a sabbatical in 2012–2013 that I spent in Cambridge, Berlin, and Bochum. Support for my research then came from a small research grant from the Wellcome Trust, from the Max Planck Institute for the History of Science in Berlin, from the Mercator Research Group at the University of Bochum, and from UCLA, my home institution. I thank colleagues in all these institutions for supporting and hosting me and providing opportunities to discuss my work. Special thanks go to Nick Hopwood, Nick Jardine, and Jim Secord at HPS in Cambridge; Christina Brandt and her colleagues in Bochum; and Veronika Lipphardt, Jenny Bangham, Samuel Coghe, João Rangel de Almeida, and Edna Suárez-Díaz at the research group Twentieth Century Histories of Knowledge about Human Variation in Berlin.

Back at UCLA, a National Science Foundation Scholars Award (#1534814, 2015–2020) provided some teaching release and support for research assistance. Additional support came from La Ville de Paris, the Institute of Advanced Studies in the Humanities in Edinburgh, and the Max Planck Institute for the History of Science in Berlin. I thank all these institutions for support, and Claudine Cohen, the late Jean Gayon, Michel Veuille in Paris, and Steve Sturdy in Edinburgh for hosting me.

My next big thank-you goes to archives and libraries that granted access to their holdings and guided my research. I would like to mention in particular Marco Fraccaro, whose immense collection of relevant papers on the history of chromosome research at the Collegio Cairoli in Pavia provided the starting point for this project. Also exceptionally helpful were Maureen Bulman at Harwell, Siobhan Marron at the MRC Human Genetics Unit in Edinburgh, Elizabeth Manners at Guy's Hospital in London, Birgitta Lindholm and Marika Hallström at the University Library in Lund, Matthias Schwerdt and Ellen Garske at the Max Planck Institute for the History of Science in Berlin, Marie Villemin at WHO, Andrew Harrison at the Medical Archives at Johns Hopkins, John Rees at the National Library of Medicine, and, last but not least, the dedicated staff at the Young Research Library at UCLA.

Visits to these and other archives would not have been as pleasant and productive without the support of local colleagues and friends who offered meals, lodging, and local know-how, and who patiently listened to my chromosome findings. I thank in particular Marco Fraccaro in Pavia; Lotte Bredt, Nicky Spice, Rachel Yates, and Jordi Prats Torné in London; Pam and Barry Barnes and Steve Sturdy in Edinburgh; Anna Thunlid and Bengt Bengtsson in Lund; Ilka Seifert, Ann Jardin, and Stefan and Bärbel Macher in Berlin; Bruno Strasser in Geneva; Simone de Chadarevian in Ottawa; Alessandra Duncan in Montreal; Susan Lindee in Philadelphia; Joanna Radin in New Haven; Julia Frank in Baltimore; and Gabriele Schilz and Drew Joseph in Washington, DC.

Equally important for the development of the project was the generous collaboration of participants who agreed to be interviewed or answered questions in writing, and who provided access to private collections. I thank in particular Pamela Barnes (Edinburgh), who was extraordinarily helpful in identifying and putting me in touch with some of her earlier colleagues; Wendy Bickmore (Edinburgh); Richard Buldock (Edinburgh); Stephen Cederbaum (UCLA); Ann Chandley (Edinburgh); Susan Collyer (Edinburgh); Alessandra Duncan (Montreal); Bernard Dutrillaux (Paris); Anthony Edwards (Cambridge); Malcolm and Marie Ferguson-Smith (Cambridge); David FitzPatrick (Edinburgh); Marco Fraccaro (Pavia); Nick Hastie (Edinburgh); Ernest Hook (Berkeley); Patricia Jacobs (Salisbury); Sibel Kantarci (UCLA); Kenneth Kidd (New Haven); Mary Lyon (Harwell); Barbara Migeon (Baltimore); Rao Nagesh (UCLA); Shirley Ratcliffe (Edenbridge); Denis Rutovitz (Edinburgh); Peter Smith (London); and John Yates (Cambridge).

Also very helpful were more informal conversations with Peter Harper (Cardiff), Shona Kerr (Edinburgh), Anne Bishop McKusick (Baltimore), Nancy Ruddle (New Haven), Carlos Sonnenschein (São Paolo), Suzanne Anker (New York), and Gina Glover (London). Peter Harper early on shared full transcripts of some of his interviews. Daniel Kevles kindly granted permission to the relevant interviews he conducted for *In the Name of Eugenics*. Nathaniel Comfort and Marsha Meldrum made Johns Hopkins–UCLA oral history interviews available. Many thanks to all of them.

Over the years, research assistance was provided by Norma Boster, Daniel Chen, Kory Fleischman, Christopher McQuilkin, Rachel Mundstock, Daniella Perry, Alex Woodman, and especially by Courtney Cruz and, most recently, Ally Osterland and Christopher Acevedo, who tracked down seemingly endless lists of books and articles. Iris Clever provided expert help with archival research and permission requests.

Several workshops and research collaborations were particularly helpful in thinking through some of the issues surrounding human heredity. The series of workshops convened under the Cultural History of Heredity project organized by Hans-Jörg Rheinberger and Staffan Müller-Wille at the (former) Department III at the Max Planck Institute for the History of Science offered a stimulating forum for reflecting on historiographical approaches to the study of heredity. The last workshop in the series, organized by Bernd Gausemeier, Staffan Müller-Wille, and Edmund Ramsden in Exeter in 2010, offered the opportunity to think in broader terms on the history of human heredity in the twentieth century. Two meetings on graphing genes, cells, and embryos—organized by Denis Thierry and Sabine Brauckmann in Naples and Berlin in 2007 and 2008—offered the very first opportunity to reflect on the visual dimensions of chromosome research. A workshop in Paris and a follow-up one co-organized with Jenny Bangham in Cambridge brought together a group of scholars interested in exploring the role of populations in studies of postwar human heredity. Talks presented at organized sessions, workshops, and seminars at UCLA and well beyond provided further opportunities for discussions and exchanges on the many aspects of the history of chromosomes. Conversations over the years with Jenny Bangham, Nathaniel Comfort, Emma Kowal, Susan Lindee, Ilana Löwy, Aryn Martin, Diane Paul, Joanna Radin, Sarah Richardson, María Jesús Santesmases, Edna Suárez-Díaz, and Alice Wexler were particularly helpful.

I received generous feedback on single or multiple chapters from Patrick Allard, Iris Clever, Alessandra Duncan, Rosemarie Garland-Thomson, Nathan Ha, Richard Henderson, Nick Hopwood, Jaehwan Hyun, Nick Jardine, Johannes Kassar, Gerald Kutcher, Lisa Onaga, Stefanie Reichelt, Jim Secord, Mary Terrall, Alice Wexler, and John Yates. Ted Porter valiantly volunteered to read the whole manuscript at an early stage. Finally, I thank the participants of the fellows' meetings at the Institute for Society and Genetics and the graduate students in the Things and Images seminar at UCLA in 2018 who helped me sharpen my arguments. More specific acknowledgments appear in the notes and figure captions.

Karen Merikangas Darling at the University of Chicago Press provided encouragement and support for this project over many years. Together with Tristan Bates and Christine Schwab she expertly shepherded the manuscript through the review and publication process. Heartfelt thanks to her and the wider team at the press who made it all happen as well as to the anonymous reviewers who helped give the manuscript its final shape. I am especially grateful to Katherine Faydash for her thoughtful copyediting, Elise Wormuth for help with the index, Shiraz Abdullahi Gallab for the cover design, and Deirdre Kennedy for all marketing aspects. Last but not least, a special thank-you to Steve at Nota Bene who stepped in at a crucial moment and saved the manuscript.

Some people heard more about chromosomes than they would normally care to. I thank Michael, Livia, and Flavia Cahn for their encouragement and distraction.

Los Angeles, January 2019

Material included in this book has previously appeared in the following publications: "Putting Human Genetics on a Solid Basis: Human Chromosome Research, 1950s–1970s," in *Human Heredity in the Twentieth Century*, edited by B. Gausemeier, S. Müller-Wille, and E. Ramsden (London: Pickering and Chatto, 2013), 141–52; "Chromosome Surveys of Human Populations: Between Epidemiology and Anthropology," *Studies in History and Philosophy of Biological and Biomedical Sciences* 47A (2014): 87–96; "Chromosome Photography and the Human Karyotype," *Historical Studies in the Natural Sciences* 45 (2015): 115–46; "Human Population Studies and the World Health Organization," *Dynamis* 35 (2015): 359–88; "'It Is Not Enough, in Order to Understand the Book of Nature, to Turn over the Pages Looking at the Pictures: Painful Though It May Be, It Will Be Necessary to Learn to Read the Text': Visual Evidence in the Life Sciences, c. 1960," in *Traces*, edited by B. Bock von Wülfingen (Berlin: De Gruyter, 2017), 55–64; "Whose Turn? Chromosome Research and the Study of the Human Genome," *Journal of the History of Biology* 51 (2018): 631–55.

Note on Sources

Many of the protagonists in this book did not consider their papers worth keeping. In a few instances, I arrived on the scene just in time to prevent more papers from being piled up for recycling. In other instances, papers were still in private hands, not yet cataloged, or fell under various restrictions, as is often the case with recent collections. Given the many gaps in the archival record, the availability of various interview collections was particularly valuable. I conducted additional interviews myself.

Interview Collections

GenMedHist. Genetics and Medicine Historical Network. Interviews recorded by Peter Harper. Transcripts available online at https://genmedhist.eshg.org/interviews/recorded-interviews/.

RAC. Rockefeller Archive Center, North Tarrytown, NY, Kevles (Daniel J.) Papers, Oral History Interview Transcripts, 1982–1984 (FA497)

OHHGP. UCLA–Johns Hopkins Oral History of Human Genetics Project. Transcripts available online at http://ohhgp.pendari.com/Collection.aspx.

If not otherwise indicated, the interviews cited in the notes are by the author.

Main Archives Visited

APSL. American Philosophical Society Library, Philadelphia

British Library, London, Archives and Manuscripts

Fraccaro Collection. Marco Fraccaro Collection, Collegio Cairoli, Pavia, Italy

Guy's Hospital, King's College London School of Medicine, Department of Medical and Molecular Genetics

Hopkins Medical Archives. The Alan Mason Chesney Medical Archives of the Johns Hopkins Medical Institutions, Baltimore

Lund University Library and Archives

MRC Human Genetics Unit, Library and Archive, Edinburgh

MRC Radiobiological Research Unit, Harwell, Library and Archive

NA. The National Archives (UK), Kew

National Archives Canada, Ottawa

NHGRI Archive. National Human Genome Research Institute, Archival and Digitized Materials, Bethesda, MD

NLM. National Library of Medicine, National Institutes of Health, Rockville, MD

Pasteur Institute, Archives, Paris

Royal Institution of Great Britain, London (Archive Collections)

Royal Society, London, Library and Archives

UCL Library Special Collections. University College London, Library Special Collections

Wellcome Library. Wellcome Library, Archives and Manuscripts, London

WHO Archives. World Health Organization Archives, Geneva

Notes

Introduction

1. Harnden, "Early Studies."

2. Penrose, introduction to "New Aspects."

3. Hsu, *Human and Mammalian Cytogenetics*, 24. Historians are rightly wary of beginnings and origin accounts that are often retrospectively created and always have their own histories. Nevertheless, here they serve to point to established accounts of postwar genetics that are themselves meaningful. The consolidation of the new chromosome count was a protracted process and as much an expression as the start of a renewed interest in human chromosome research; see de Chadarevian, "Chromosome Photography."

4. Hsu, *Human and Mammalian Cytogenetics*; Lima-de-Faria, *One Hundred Years*; Harper, *First Years*.

5. On the impact of the new generation of light microscopes on cell biology, including the observation of chromosomes, in the 1880s, see Coleman, *Biology in the Nineteenth Century*, 22–41. For a visual rendering of the "dance of chromosomes," see the illustrations in Auerbach, *Genetics in the Atomic Age*. In her brief tract, Charlotte Auerbach, a scientist and activist working in Edinburgh, aimed to introduce a general public to the principles of genetics to help people understand the deleterious effects of atomic radiation. On the publication, including the illustrations, and Auerbach's role as a female public scientist, see Richmond, "Women as Public Scientists."

6. For examples, see Darlington and Ammal, *Chromosome Atlas*; Makino, *Atlas*. For a later multivolume example in this same tradition, see Benirschke and Hsu, *Atlas*. See also figure 4.9 on p. 134 in this volume.

7. Painter, "Comparative Study." On speculations about different chromosome numbers in different groups of people, see chapter 4 in this volume.

8. Hsu, *Human and Mammalian Cytogenetics*, 8. With new sterilization laws that were put into place in various American states in the 1920, testes material

became somewhat more readily available. Tissue of women amenable for chromosome counts remained difficult to obtain; see Kevles, *In the Name of Eugenics*, 238–39.

9. Harman, *Man Who Invented the Chromosome*; Santesmases, "Cereals, Chromosomes and Colchicine"; Campos, *Radium*; Curry, *Evolution Made to Order*.

10. Kohler, *Lords of the Fly*.

11. Beatty, "Genetics in the Atomic Age"; Beatty, "Scientific Collaboration"; Lindee, *Suffering Made Real*; Rader, *Making Mice*; de Chadarevian, "Mice and the Reactor"; Creager, *Life Atomic*.

12. On the recount, see Kottler, "From 48 to 46"; Martin, "Can't Any Body Count?"; Harper, "Discovery of the Human Chromosome Number"; de Chadarevian, "Chromosome Photography."

13. Ferguson-Smith, "From Chromosome Number to Chromosome Map."

14. For richly detailed introductions to the history of the field, often informed by firsthand knowledge, see Hsu, *Human and Mammalian Cytogenetics*; Harris, *Cells of the Body*; Harper, *First Years*; as well as the historical introduction to Vogel and Motulsky, *Human Genetics*. Although historically cytogeneticists focused their attention on the study of chromosomes in the cell nucleus, other cell organelles, such as mitochondria, also include genetic elements. In addition, cytoplasmatic mechanisms not involving genes also occur in the cell. For a historical account of these research traditions and the controversies surrounding them, see Sapp, *Beyond the Gene*.

15. On a similar approach of following specific scientific objects and using them as historical "tracers," see, for example, Lynch et al., *Truth Machine*; and Creager, *Life Atomic*.

16. Penrose, introduction to "New Aspects," 3.

17. World Health Organization, *Research in Population Genetics of Primitive Groups*; Weiner and Lourie, *Human Biology*.

18. Mazumdar, *Eugenics, Human Genetics and Human Failings*; Kevles, *In the Name of Eugenics*, 213; Schneider, "Blood Group Research"; Schneider, "History of Research on Blood Group Genetics"; Gannett and Griesemer, "ABO Blood Groups"; Bangham, "Blood Groups and the Rise of Human Genetics"; Bangham, "Blood Groups and Human Groups."

19. On the study of human heredity and its inextricable link to eugenics, see Mazumdar, *Eugenics, Human Genetics and Human Failings*; Kevles, *In the Name of Eugenics*; Paul, *Controlling Human Heredity*; Comfort, *Science of Human Perfection*. On the eugenic movement, its development in various countries and the persistence of eugenic practices more generally, see, for example, Duster, *Backdoor to Eugenics*; Stern, *Eugenic Nation*; Mazumdar, *The Eugenics Movement*; Cottebrune, *Der planbare Mensch*; Schmuhl, *Kaiser Wilhelm Institute*; Bashford and Levine, *Oxford Handbook*; Bashford, "Epilogue"; Largent,

Breeding Contempt; Lombardo, *Century of Eugenics*; Turda and Gillette, *Latin Eugenics*. For recent attempts to broaden the questions around human heredity, see Lindee, *Moments of Truth*; Gausemeier, Müller-Wille, and Ramsden, *Human Heredity*; Bangham and de Chadarevian, "Special Section."

20. Hubbard, "Abortion and Disability."

21. L. Penrose, "Human chromosomes" (typescript for a lecture), 22 October 1959, p. 5, file 88/1, Penrose Papers, University College London (UCL) Library, Special Collections. Some of the studies on human heredity already mentioned dedicate a chapter or part of a chapter to human chromosome research. Although pioneering in their kind, the discussion remains focused on medical genetics and eugenics; see, for example, Kevles, *In the Name of Eugenics*, chap. 16; Comfort, *Science of Human Perfection*, chap. 6. Other historical accounts of postwar human genetics simply skip over relevant developments in human chromosome research. See, for example, the discussion of human gene mapping in Pauline Mazumdar's otherwise excellent book, *Eugenics, Human Genetics and Human Failings*. In their recent attempt of a *longue durée* cultural history of heredity, Staffan Müller-Wille and Hans-Jörg Rheinberger devote much attention to the establishment and broad acceptance of the chromosomal theory of heredity in the early twentieth century but then pass on to consider the changes in the concept of heredity brought about by molecular biology and especially genetic engineering and current genomic practices; Müller-Wille and Rheinberger, *Cultural History of Heredity*. Recently, some historians have started paying more attention to the history of human chromosomes, especially with respect to the clinic and to biological theories of sex. Together they make the case for the importance of human chromosome research in postwar biology and medicine and in the cultural history of heredity. For clinical studies, see Gaudillière, "Le syndrome nataliste"; Gaudillière, "Whose Work Shall We Trust?"; Hogan, *Life Histories*; Hopkins, "Hidden Research System"; Lindee, *Moments of Truth*; Löwy, "How Genetics Came to the Unborn"; Löwy, *Imperfect Pregnancies*; Santesmases, "Human Autonomous Karyotype"; Santesmases, "Human Chromosomes and Cancer." On the history of sex chromosomes, see Ha, "Marking Bodies"; Richardson, *Sex Itself.* For recent historical studies on cell biology more generally, see Landecker, *Culturing Life*; O'Malley and Müller-Wille, "Cell as Nexus"; Santesmases and Suárez-Díaz, "Cell-Based Epistemology"; Matlin, Maienschein, and Laublicher, *Visions of Cell Biology.*

22. Human somatic cells have two sets of twenty-three chromosomes (one set from each parent). Sperm and egg cells carry only one chromosome set.

23. The description of chromosomes presented here is indebted to a number of approaches that have been proposed to deal with the technical, epistemic, visual and political dimensions of scientific objects and the conditions under which they come into being and become amenable to scientific study. These include especially Hans-Jörg Rheinberger's notion of "epistemic things" (Rheinberger,

Toward a History), studies on the "biography of things" (Appadurai, *Social Life of Things*; Henare, Holbraad, and Wastell, *Thinking through Things*; Daston, *Things That Talk*) and on images and visuality in scientific practices (Hacking, *Representing and Intervening*; Lynch and Woolgar, *Representation in Scientific Practice*; Daston and Galison, *Objectivity*; Nasim, *Observing by Hand*; Hopwood, *Haeckel's Embryos*). Also relevant is the notion of biomedical platforms that sustain specific scientific entities and routines; see Keating and Cambrosio, *Biomedical Platforms*. On the visual dimensions of chromosome analysis, see also Turrini, "Continuous Grey Scales"; Santesmases, "Biological Landscape of Polyploidy"; Santesmases, "Circulating Biomedical Images"; Hogan, *Life Histories*.

24. de Chadarevian, "Chromosome Photography." For Levan an advantage of drawing was that even "non-photogenic cells" were useful for analysis. In contrast, photographic analysis depended on the "always exceptional [i.e., rare] photogenic cells"; see Albert Levan, "Conference on Human Chromosomes—Colorado, April 8–11, 1960: Comments on points A-G of the provisional agenda," p. 6, file Normal karyotype, Marco Fraccaro Collection, Collegio Cairoli, Pavia, Italy. On the presence of aesthetic judgment in all stages of chromosome analysis, see Turrini, "Continuous Grey Scales." On the epistemic connection between seeing and drawing in the field of astronomy, see Nasim, *Observing by Hand*. On the politics and ethics of representation in photography and the tension between knowing from photography and understanding, see Sontag, *On Photography*, 23.

25. Hacking, *Representing and Intervening*, 186.

26. On the "boundary status" of microscopic slides as both "highly artificial" and "natural objects," see Löwy, "Microscope Slides."

27. Rheinberger, *Epistemology of the Concrete*, 218 and 243.

28. Rheinberger, *Epistemology of the Concrete*, 238. Despite the flattening of the specimen it is still possible to play with the focus to gauge the depth of specific structures. This exactly marks the difference between looking at a specimen under the microscope and at a photographic image of the viewing field. On the essential flatness of the microscopic view and its connection to the management of living matter, see also Cartwright, *Screening the Body*, 90–91.

29. Rheinberger, *Epistemology of the Concrete*, 219.

30. For epistemic considerations around the practice of microscopy and its history, see also Hacking, *Representing and Intervening*, 186–209; and Schickore, *Microscope and the Eye*.

31. Martin, "Can't Any Body Count?"

32. On the cutting of chromosomes as a gendered activity, see Drucker, "Janet Rowley"; Santesmases, "Circulating Biomedical Images," 406–8. On cutting and pasting as a technique in science and art, see te Heesen, *Newspaper Clipping*. Small envelopes with cutout chromosomes as well as single loose chro-

mosomes that have fallen off from cut-and-paste karyotypes are a frequent find in archival collections of chromosome researchers.

33. Hsu, *Human and Mammalian Cytogenetics*, 127.

34. The descriptions of current cytogenetic practices are based on first-hand observation in a modern cytogenetic laboratory performing clinical diagnoses. Despite the increasing role of automation the training period for a cytogeneticist has not changed much since the 1950s and chromosome analysis remains a highly skilled activity that attracts dedicated practitioners. On the automation of chromosome analysis, see chapter 4 in this volume.

35. Daston, "On Scientific Observation," 107.

36. Daston, "On Scientific Observation," 102. On observation as a skill, based on training and practice, see also Fleck, *Genesis*, 84–98, and Hacking, *Representing and Intervening*, 168; on the engagement of the practitioner's body in scientific observation, see Rasmussen, *Picture Control*; on the history of scientific observation, see Lunbeck and Daston, *Histories of Scientific Observation*.

37. Daston, "On Scientific Observation," 107–8.

38. Hsu, *Human and Mammalian Cytogenetics*, 5.

39. Alessandra Duncan, personal communication, email, 16 April and 10 June 2014.

40. The technique, developed in the early 1990s, uses fluorescent molecular probes to tag specific chromosomal sites. The result can be viewed under a fluorescence microscope and in practice allows for a molecular resolution of the chromosome image. See chapter 5 in this volume.

41. Kevles, *In the Name of Eugenics*, 205–11; Zallen, "Medical Genetics in Britain."

42. Clarke and Fujimura, *Right Tools for the Job*.

Chapter One

1. See, for example, Hsu, *Human and Mammalian Cytogenetics*; and Lima-de-Faria, *One Hundred Years*. On pre-banding chromosome studies as "paleolithic," see Lejeune, "Scientific Impact," 24.

2. R. B. Scott, "The Treatment of Leukaemia" (unpublished lecture), p. 2, file PP/RBS/C41, folder "Unpublished papers 1958–1960," Ronald Bodley Scott Collection, Wellcome Library Archives and Manuscripts, London (also quoted in Kraft, "Manhattan Transfer," 209).

3. On the charged notion of a permissible dose in radiation protection, see Walker, *Permissible Dose*.

4. The person credited with having coined the term is the *New York Times* journalist William L. Laurence. Having witnessed the first atomic test explosion in the New Mexico desert in July 1945, he declared this to be the beginning of

the atomic age in his first report of the event a few months later; see Laurence, "Drama of the Atomic Bomb." A slightly earlier piece in the *New York Times Magazine* also used the phrase; see Davis, "We Enter a New Era." Laurence was not named as an author, but he was the driving force in the reporting on the Manhattan Project at the *New York Times*. Despite the harrowing images that soon started to emerge from Hiroshima and Nagasaki, the phrase "atomic age" was mainly used to denote a feeling of nuclear optimism prevalent in the 1950s and certainly shared by Laurence (after first winning the Pulitzer Prize for his atomic reporting he later became accused of partial reporting of the Japanese bombing events). In the 1960s the term lost currency but survived especially in pop culture, where it gained increasingly threatening connotations. I thank Alex Wellerstein for information on Laurence. On Laurence and the *New York Times* reporting on the atomic bomb, see also Deepe Keever, *News Zero*.

5. Roff, *Hotspots*; Gordin, *Five Days in August*.

6. In his monumental history *Radioactivity and Health*, Newell Stannard argued that, contrary to what has often been stated, the possibility of nuclear fallout was seriously considered by the scientists of the Manhattan Project, especially in relation to the Trinity test. In the context of the atomic bombing on Japan fallout received less attention. What was not considered in either case at the time were the long-term effects of radioactive fallout; see Stannard, *Radioactivity and Health*, 879. On the extensive radiobiological program of the Manhattan Project that was aimed at keeping the people involved in developing the radioactive technology safe, see Hacker, *Dragon's Tail*.

7. The five prints first appeared in the Japanese magazine *Asahi Gurafu* in August 1952. One month later *Life* magazine ran a report with Matsushige's and other previously censured photographs; see "When Atom Bomb Struck."

8. Beatty, "Genetics in the Atomic Age"; Beatty, "Scientific Collaboration."

9. Lindee, *Suffering Made Real*, 192.

10. Neel and Schull, *Effect of Exposure*, 204.

11. Concerns about the effects of radiation on the human genome clearly built on earlier eugenic preoccupations with the "genetic load" of human populations; see Paul, "'Our Load of Mutations' Revisited." Without reference to these earlier anxieties it is not possible to explain why genetic effects—above any other effect—came to dominate the research agenda in the 1940s and 1950s. Nevertheless, it is important to recognize the fundamental differences. Genetic concerns in the postwar era focused on increased cancer risks and disease burden (not on the genetic worth of different groups of people), and the problems concerned all people, independent of social class and ethnic affiliation. Also, genetic risks included somatic and congenital (present at birth but not necessarily inherited) next to hereditary mutations in the germ plasma. Chromosome analysis became a tool to study all these effects.

12. Creager, "Mutation." In accordance with other studies, Creager attributes the original separation of the two issues to the professional separation of toxicologists and pathologists who were responsible to study the somatic effects of radiation from geneticists who studied the genetic effects. In contrast to geneticists, toxicologists postulated a safe threshold for radiation exposure. With the gradual acceptance of the somatic mutation theory of carcinogenesis from the late 1950s, this distinction started to break down. On the debate about the somatic effects of radiation and research on chromosomal aberrations, see also Semendeferi, "Legitimating a Nuclear Critic."

13. de Chadarevian, "Mice and the Reactor"; Rader, *Making Mice.*

14. E. Mellanby to A. Barlow, 27 September 1946, file FDI 468, National Archives (UK), Kew (hereafter NA).

15. Thomson, *Origins and Policy*, 104; Thomson, *Programme of the Medical Research Council*, 58–65 and 104–6.

16. Hacker, *Elements of Controversy*, 199 and 254.

17. On the mice experiments at Harwell and Oak Ridge, see de Chadarevian, "Mice and the Reactor"; and Rader, *Making Mice.*

18. Mutations also played a central role as both object and tool of research in the early history of molecular genetics. Not by chance, Crick and Watson stressed that their model of the structure of DNA offered a molecular explanation for mutational events; Watson and Crick, "Genetical Implications," 966. Mutants, mutations, and mutagens played a crucial role in establishing protein-DNA relationships and the basic features as well as universal character of the genetic code. Just around the time when human chromosome research was taking off, molecular biologists linked changes in amino acid sequences of proteins to mutations at the DNA level; see Ingram, "Specific Chemical Difference"; Ingram, "Gene Mutations in Human Haemoglobin." From the 1960s, electrophoresis, protein fingerprinting, and, eventually, protein sequencing were widely used for identifying protein variants, for diagnostic purposes (e.g., for identifying sickle cell hemoglobin), for genetic studies of populations and for building evolutionary trees. The sequencing of DNA did not seem attainable at the time.

19. Kevles, *In the Name of Eugenics*, 205.

20. On Ford's career, see Lyon, "Charles Edmund Ford"; interview with Charles Ford by Daniel Kevles, Abbington, 25 June 1982, Rockefeller Archive Center (RAC), North Tarrytown, NY, Kevles Papers, FA497, box 1, folder 11; and interview with John Evans by Peter Harper, Edinburgh, 10 December 2003, Genetics and Medicine Historical Network recorded interviews (GenMedHist). The addition of a cytogeneticist to the MRC group at Harwell had been a desideratum from the very beginning, but finding one did not prove easy. Ford's name was first mentioned to Himsworth by John Cockcroft, the director of the

Atomic Energy Research Establishment, who knew of Ford through his previous appointment as director of the Chalk River Laboratory in Canada; see E. Mellanby (notes on discussion with J. Cockcroft), 15 November 1946, file FD 1/468, NA.

21. On the special skill involved in squashing and the original mistrust against the import of such a crude "horticultural" method into work with mammalian tissues, see Harper, *First Years*, 11–12.

22. On the intricate development of bone marrow transplantation from research tool to an adjunct to cancer treatment in the Cold War context as well as on the Cold War origins of stem cell research, see Kutcher, *Contested Medicine*; and Kraft, "Manhattan Transfer."

23. Tjio and Levan, "Chromosome Number of Man."

24. At this point, Levan and Tjio did not exclude the possibility that germ cells had forty-eight chromosomes after all but other cells in the body might lose a few; see "A Life with Chromosomes: An Interview with Albert Levan, Professor of Cytology, by Professor Bengt Olle Bengtsson, Professor of Genetics, Lund University 1989," Dialogue Project, Lund University. A Swedish version of the filmed interview is available at http://www.alvin-portal.org/alvin/view.jsf?pid =alvin-record:275185. I thank Bengtsson for making a version with English subtitles available to me.

25. C. Ford to A. Levan, 30 July 1956, Correspondence folders, Albert Levan Papers, Special Collections, Lund University Library and Archives. See also C. E. Ford, "Human Cytogenetics."

26. de Chadarevian, "Chromosome Photography," 131–33; and pp. 134–36 in this volume.

27. Among these were John Hamerton, Patricia Jacobs, David Harnden, and Ted Evans. Janet Rowley, visiting from the University of Chicago, who later made decisive contributions to cancer cytogenetics, intended to study under him, but Ford could not take her on at that moment; see interview with Janet Rowley, June 2005, transcript, UCLA–Johns Hopkins Oral History of Human Genetics Project (OHHGP).

28. "Extract from letter dated 25.1.54 from Dr. W. M. Court Brown to Sir Harold Himsworth," file FD 23/1310, NA.

29. [F. H. K. Green?] (MRC) to W. M. Court Brown, 31 December 1954, and [F. H. K. Green?] to T. F. Fox (editor of *The Lancet*), 19 January 1955, file FD 23/1310, NA.

30. F. H. K. Green (MRC) to Prof. D. W. Smithers (Radiotherapy Department, Royal Marsden Hospital, London), 25 June 1955, FD 23/1310, NA.

31. W. M. Court Brown to F. H. K. Green (MRC), 5 April 1955, FD 23/1310, NA. A brief letter, personally signed by Himsworth, calling on readers to communicate any case of leukemia in spondylitis patients they may be aware of, appeared in the *Lancet* and the *British Medical Journal* in January 1955; Hims-

worth, "Leukaemia." For the first preliminary report on the investigations, see Court Brown and Abbatt, "Incidence of Leukaemia."

32. [F. H. K. Green] to W. M. Court Brown, 19 April 1955, FD 23/1310, NA.

33. Smith, "1957 MRC Report on Leukaemia," B6.

34. Darby, "Conversation with Sir Richard Doll."

35. Court Brown and Doll, "Appendix B"; Court Brown and Doll, *Leukaemia and Aplastic Anaemia.*

36. Court Brown and Doll, "Appendix A."

37. The study marked the beginning of a long collaboration between the two researchers. Their common work included studies on the cancer mortality of British radiologists and on the incidence of leukemia after diagnostic exposure in the uterus. In 1959 Court Brown and Doll traveled to Japan to study firsthand the leukemia incidence in atomic bomb survivors. Alice Stewart at the MRC Social Medicine Research Unit at Oxford also performed studies on the increased incidence of leukemia in children born from mothers who had undergone X-ray analysis during pregnancy. However, her studies found recognition only later. On the controversies surrounding Stewart's work, see Greene, *Woman Who Knew Too Much.* In later years Doll, originally one of her critics, fully acknowledged Stewart's contributions; see Richard Doll, "Radiation Hazards: 25 Years of Collaborative Research: The Silvanus Thompson Memorial Lecture given at the Royal Free School of Medicine on April 17th, 1980" (typescript), file PP/DOL/B/1/12, Richard Doll Papers, Wellcome Library.

38. Interview with John Evans by Peter Harper, Edinburgh, 10 December 2003, GenMedHist.

39. See [W. M. Court Brown], "Group for Research into the General Effects of Radiation: Progress Report" [1960]; MRC Harwell, Library and Archive. I thank Peter Harper for kindly providing me with a copy of this report.

40. Ford, Jacobs, and Lajtha, "Human Somatic Chromosomes."

41. Medical Research Council, *Hazards to Man*; Medical Research Council, *Hazards to Man: Second Report.*

42. W. M. Court Brown, "The Effects of Radiation on the Health of the Individual: Summary of Points from the 1956 Report Which Require Comment in the Proposed Supplementary Report," MRC 59/954, November 1959, file F 241, Joseph Mitchell Papers, Cambridge University Library.

43. de Chadarevian, "Mice and the Reactor," 722–27.

44. Medical Research Council, *Hazards to Man*, 74–75.

45. T. Puck, "Why Low-Level Human Radiation Damage Is Difficult to Assess, but Dangerous to Ignore" (undated), uncataloged file box, Theodore T. Puck Collection, University of Denver, Penrose Library Special Collections and Archives, Denver, CO. I thank Daniella Perry for pointing me to this document.

46. L. S. Penrose to L. H. Gray, 1 April 1955, Penrose Papers, file 79E Correspondence, UCL Library Special Collections.

47. L. S. Penrose, "Memorandum on Gene Mutation in Man" (typescript, draft), 10 May 1955, Penrose Papers, file 79F, UCL Library Special Collections.

48. Interview with Ursula Mittwoch by Peter Harper, London, 2 March 2004, GenMedHist.

49. G. Pontecorvo to P. C. Koller, 4 April 1955, and G. Pontecorvo to L. S. Penrose, 5 April 1955, Penrose Papers, file 79E, folder 2, UCL Library Special Collections.

50. Minutes of meeting, 7 July 1955 (handwritten by Penrose), file GS/3/2, Minute book of Committee meetings, 1952–1968, Genetics Society Archives, John Innes Foundation Historical Collection, Norwich, UK.

51. Medical Research Council, *Hazards to Man*, appendixes C, D and E.

52. "Human Chromosomes" (typescript for a lecture), 22 October 1959, p. 5, file 88/1, Penrose Papers, UCL Library Special Collections. On Penrose's disappointment at failing to find the trisomy in Down syndrome, see L. S. Penrose to J. B. S. Haldane, 4 June 1959, file 136, Penrose Papers, UCL Library Special Collections, as well as interview with Marco Fraccaro, 30 March 2004, Pavia (Italy). On Penrose and Down, see also chapter 2 in this volume.

53. Harper, *First Years*, 17–18; Harper, *Short Story*, 428–53. On Trofim Lysenko's influence on Soviet genetics, see Krementsov, *Stalinist Science*; Roll-Hansen, *Lysenko Effect*; and Gordin, "Lysenko Unemployed."

54. Robinson, "Living History," 478.

55. On the reestablishment of medical genetics, including cytogenetics, in the post-Lysenko era, see S. Bauer, "Mutations in Soviet Public Health Science." On the Denver conference, see chapter 2 in this volume.

56. On the cytogenetic tradition at Svalöf, see Ellerström and Hagberg, "Cyto-Genetic Department." The Knut and Alice Wallenberg Foundation funded both the creation of the Cytogenetics Department at Svalöf and the Laboratory for Cell Research at the Karolinska Institute in Stockholm, headed by Torbjörn Caspersson. Both laboratories made decisive contributions to cytogenetics.

57. On Levan's early work on cancer chromosomes, see Santesmases, "Human Chromosomes and Cancer."

58. The strong presence of Swedish researchers in human chromosome research in the 1950s might be explained by the strong tradition of chromosome research at the plant breeding station in Svalöf where many geneticists started their career, combined with the equally strong interest in human genetics, including race genetics. Jan A. Böök, who was present in Denver, chaired the Institute for Medical Genetics (previously the Institute for Race Genetics) in Uppsala. First directed by eugenicist and Nazi sympathizer Herman Lundborg (1922–1935), the institute changed direction under his successor, Gunnar Dahlberg, author of a critical appraisal of race genetics; see Dahlberg, *Race, Reason and Rubbish*. Böök introduced cytogenetics in the institute in the late 1950s.

Marco Fraccaro, a visitor from Italy in Böök's laboratory, also attended the Denver meeting. A couple of years later, he would head a new cytogenetics laboratory in Pavia, funded by the European Atomic Energy Community (Euratom).

59. Partly on the basis of an interview with Lejeune's wife, Gaudillière presents the introduction of cytogenetic techniques in Turpin's laboratory as part of the broader radiobiological project. Although plausible and fully fitting with the broader argument of this chapter, this explanation is not directly supported by other sources, above all Gautier's memoir; see Gaudillière, "Whose Work Shall We Trust?" 74; Gautier and Harper, "Fiftieth Anniversary of Trisomy 21."

60. In a retrospective account of his research career, Chu provided a vivid account of the close networks and exchange of technical skills between researchers interested in human chromosome research and somatic cell genetics before and in the years after the Denver conference; see Chu, "Early Days of Mammalian Somatic Cell Genetics."

61. In fact, Soviet researchers had already described a method for the use of white blood cells for chromosome studies in the mid-1930s. Yet the method was not taken up at the time and was later forgotten. The well-respected Soviet group stopped publishing shortly thereafter, reflecting the politicization and rising dominance of Lysenko's influence in Soviet genetics; see Hungerford, "Some Early Studies"; Harper, *First Years*, 139–40. Nevertheless, the story can also be read as a confirmation that human chromosome analysis was less of a concern in the 1930s than in the 1960s, when researchers around the world eagerly took up the peripheral blood method. The Soviet method did not make use of phytohemagglutinin.

62. Harper, *First Years*, 8 and 15; Harper, *Short Story*, 248. For a general introduction to work on human chromosomes with special attention to the contributions by Japanese researchers and on Japanese people exposed to radiation, see Makino, *Human Chromosomes*.

63. National Academy of Sciences, *Biological Effects of Atomic Radiation*; Hamblin, "'Dispassionate and Objective Effort.'"

64. Interview with Marco Fraccaro by Peter Harper, Pavia, 21 April 2004, GenMedHist.

65. T. Puck, "Why Low-Level Human Radiation Damage Is Difficult to Assess, but Dangerous to Ignore" (undated), uncataloged file box, Theodore T. Puck Collection, University of Denver, Penrose Library Special Collections, Denver, CO.

66. Strickland, *Politics, Science, and Dread Disease*; Leopold, *Under the Radar*; Kutcher, *Contested Medicine*.

67. Lenoir and Hays, "Manhattan Project"; Kraft, "Between Medicine and Industry"; Herran, "Spreading Nucleonics"; Herran, "Isotope Networks"; Creager, *Life Atomic*.

68. Krige, "Atoms for Peace."

69. The singular focus on the association of radiation with leukemia partly obscured the connection of low-level radiation exposure with other late-onset cancers. Eventually these were regarded as an even bigger concern; see Semendeferi, "Legitimating a Nuclear Critic."

70. "Come Back to Me Again, Sadako." A letter from Sadako's mother, Fujiko Sasaki, 1956, published at *The Global Human: Social Comments* (blog), http://theglobalhuman.wordpress.com/come-back-to-me-again-sadako/.

71. Research more generally focused on the etiology as well as the treatment of leukemia, stimulated by encouraging results regarding the use of chemotherapeutic drugs; see R. B. Scott, "The Treatment of Leukaemia" (unpublished lecture), file PP/RBS/C41, folder "Unpublished papers 1958–1960," Ronald Bodley Scott Collection, Wellcome Library. The American Leukemia Research Fund was created in 1946, in line with other research organizations for cancer and a more general reorientation of existing cancer funds toward research. The British counterpart, which funded considerable research in cytogenetics, was founded in 1960; for a brief history, see Piller, *Rays of Hope.*

72. Interview with Patricia Jacobs by Peter Harper, Salisbury, 13 February 2004, GenMedHist. Jacobs's first project was to establish whether there was a chromosomal difference between spontaneous and radiation-induced human leukemias.

73. Nowell and Hungerford, "Chromosome Studies." Leukemias were classified as chronic or acute and according to the cells that were affected. Myeloid leukemia affected marrow cells that developed into white blood cells known as granulocytes, whereas lymphocytic leukemia affected marrow cells that developed into white blood cells known as leukocytes. On further developments in the classification of leukemias based on clinical and cytological observations, and its role in guiding research in cancer genetics, see Keating and Cambrosio, "New Genetics and Cancer," 343–44.

74. Harper, *First Years*, 123–25.

75. The difficulty of identifying the chromosome was compounded by the fact that the Philadelphia chromosome was found to be chromosome 21, whereas the chromosome implicated in Down syndrome was the slightly smaller chromosome 22. In order to not further confuse things, it was decided to stick to the established link between chromosome 21 and Down syndrome and to identify the Philadelphia chromosome as chromosome 22; see Ferguson-Smith, "From Chromosome Number to Chromosome Map."

76. W. M. Court Brown, "An Account of the Development of Research in the Medical Research Council's Clinical and Population Cytogenetics Research Unit in Human Cytogenetics and Related Subjects (1956–1969), with Some Observations on Likely Future Progress" (typescript), December 1968, MRC Human Genetics Unit, Library and Archive, Edinburgh.

77. de Chadarevian, *Designs for Life*, 350–52.

78. In a critical account of the "new genetics," Peter Keating and Alberto Cambrosio highlight the (often forgotten) impact of cytogenetics (and of clinical medicine and observational science more generally) on the development of the oncogene theory of cancer and its continuing importance in the field of cancer genetics; see Keating and Cambrosio, "New Genetics and Cancer." On the excitement among cancer researchers about the findings around the Philadelphia chromosome, see Christie and Tansey, *Leukaemia*, 32, 37, and 39–40.

79. On the special research effort around leukemia, and in particular leukemia viruses, see Scheffler, "Managing the Future"; Scheffler, *Contagious Cause*.

80. The fire at Windscale was the worst reactor accident at the time and remained so until the Three Mile Island accident in 1979, which, though, released much less radioactive iodine, cesium, and strontium (but considerably more xenon) into the atmosphere. The Oak Ridge accident seldom appears in these statistics that include the Chernobyl and Fukushima disasters. On the Windscale accident, see Arnold, *Windscale 1957*.

81. Initially, the Edinburgh group highly regarded the work done by Michael A. Bender and his colleagues at Oak Ridge. However, it appears that the group grew increasingly frustrated with Bender's failure to recognize some basic features of the blood culture system, especially as relating to the dynamics of the system in irradiated subjects; see W. M. Court Brown, "Memorandum on Cytogenetic Studies in Persons Exposed in Criticality and Other Radiation Accidents" (undated, enclosed with letter by W. M. Court Brown to H. Himsworth, 8 May 1963); and W. M. Court Brown to R. C. Norton, 13 September 1963, file FD 23/218, NA. At the symposium on human radiation cytogenetics, held in Edinburgh in 1966, where some of these issues were discussed, the two invitees from Oak Ridge were unable to attend, possibly for entirely unrelated reasons; see Evans, Court Brown, and McLean, *Human Radiation Cytogenetics*. Tensions between the Biological Division at Oak Ridge and the MRC unit at Harwell existed in relation to other aspects of radiobiological research, but there is no indication that such tensions affected the cytogenetic work; see de Chadarevian, "Mice and the Reactor," 727–28. By the late 1960s, the WHO promoted its program in radiation cytogenetics as an endeavor transcending Cold War tensions between East and West, with the Edinburgh unit being designated as the WHO International Reference Center for Chromosome Aberration Evaluation in Populations; see H. J. Evans to R. C. Norton, 15 September 1970, and other material in file FD 12/459, NA.

82. W. M. Court Brown to H. Himsworth, 2 June 1964, file FD 23/769, NA.

83. W. M. Court Brown to H. Himsworth, 2 June 1964, file FD 23/769, NA.

84. H. Himsworth to W. M. Court Brown, 5 June 1964, file FD 23/769, NA. *Rad* designates the absorbed dose while *rem* indicates the biologically effective

dose. For practical reasons one rad can be converted into one rem. The maximum permissible occupational dose at the time was fixed at five rems per year for whole body exposure.

85. W. M. Court Brown to H. Himsworth, 8 October 1964, file FD 23/769, NA. It should be noted that by this time Charlotte Auerbach, working with flies at the Institute of Animal Genetics in Edinburgh, had already established that mustard gas produced permanent chromosome changes. Rachel Carson, in her book *Silent Spring*, also suggested that a range of other chemicals, including pesticides, could lead to chromosome damage. Besides having herself a background in genetics, she consulted on the matter with the German émigré geneticist Klaus Patau at the University of Wisconsin, who had described trisomy 13; see Carson, *Silent Spring*, 185–92, 205–7; Richmond, "Women as Public Scientists," 371–78. Both women played important roles as public scientists in the 1950s and 1960s; see Richmond, "Women as Public Scientists."

86. W. M. Court Brown to H. Himsworth, 26 November 1964, file FD 23/769 (Report on Chromosome Studies on Subjects Accidentally Exposed at Dounreay on 12 April 1964 and Windscale—Incidence of Leukemia in Benzene Workers), NA. On the efforts to automate karyotyping, see chapter 4 in this volume.

87. Tough and Court Brown, "Chromosome Aberrations."

88. Dolphin and Lloyd, "Significance of Radiation-Induced Chromosome Abnormalities," 186. For a later evaluation of chromosome dosimetry in radiological protection, see Lloyd and Purrott, "Chromosome Aberration Analysis."

89. W. M. Court Brown to H. Himsworth, 16 November 1964, file FD 23/769, NA.

90. Bloom et al., "Cytogenetic Investigation"; Bloom et al., "Leukocyte Chromosome Studies"; Awa, Bloom, et al., "Cytogenetic Study." The cytogenetic studies of the survivors and their offspring continued for many decades. For a report on the later studies, see Awa, Honda, et al., "Cytogenetic Study."

91. Evans, Court Brown, and McLean, *Human Radiation Cytogenetics*.

92. Evans et al., "Radiation-Induced Chromosome Aberrations." See also typescript with chronological listing of research highlights at the Edinburgh unit (no title; handwritten note on top reads: H. J. E[vans]), 6 December 1993, MRC Human Genetics Unit, Library and Archive, Edinburgh. The population was too small to establish any possible correlation with an increased incidence of cancer in the same population, yet such studies were initiated by the NRPB.

Chapter Two

1. Harper, *First Years*, 77. The quote is attributed to Paul Polani. The spirit, if not the exact quote, is reflected in Paul Polani, "The year is 1959 . . ." (2003,

unpublished autobiographical memoir); Paul Polani, Personal Information file, Royal Society Library and Archives.

2. Lejeune, Gautier, and Turpin, "Les chromosomes humains." A congenital condition is present at birth. The condition can be inherited or acquired during development. Trisomy 21 is an example of a congenital condition that is genetic but not inherited. In a historical review article, Polani listed the 1958 confirmation by Ford, Jacobs, and Lajtha of a female sex complement in a male with Klinefelter syndrome as "the very first human developmental chromosome anomaly ever documented chromosomally and reported"; see Polani, "Human and Clinical Cytogenetics," 120. Perhaps because the chromosomal observation merely confirmed what the Barr body test had already suggested, or perhaps because later observations showed that Klinefelter syndrome in fact is characterized by an XXY chromosome set, the report by the Paris group on Down syndrome is generally regarded as the first confirmed case of a congenital chromosome anomaly. With hindsight the suggestion has also been made that the supposed Klinefelter case described in the 1958 paper might have been clinically misdiagnosed and in reality was a sex-reversed XX male; see Harper, *First Years*, 84. On the Barr body test, see pp. 47–48 in this volume.

3. Jacobs and Strong, "Case of Human Intersexuality."

4. Lejeune, Gautier, and Turpin, "Étude des chromosomes"; Lejeune, Turpin, and Gautier, "Le mongolisme."

5. C. E. Ford et al., "Sex-Chromosome Anomaly"; C. E. Ford, et al., "Chromosomes in a Patient"; Jacobs et al., "Somatic Chromosomes"; Jacobs et al., "Evidence."

6. Harnden, "Early Studies," 165; "Chromosomes of Man" (editorial).

7. Lennox, "Chromosomes for Beginners," 1046.

8. "Human Chromosomes" (typescript for a lecture), 22 October 1959, file 88/1, Penrose Papers, UCL Library Special Collections. Most likely Penrose here referred to the chromosome picture of the Klinefelter's Down patient that Charles Ford had prepared after receiving some bone marrow cells of the patient from him.

9. López-Beltrán, "Medical Origins of Heredity."

10. T. Puck, "Why Low-Level Human Radiation Damage Is Difficult to Assess, but Dangerous to Ignore" (undated), uncataloged file box, Theodore T. Puck Collection, University of Denver, Penrose Library Special Collections, Denver, CO.

11. On the construction of genetic disease categories, see Yoxen, "Giving Life a New Meaning"; Lindee, "Genetic Disease"; Comfort, *Science of Human Perfection*; Hogan, *Life Histories*; Navon, *Mobilizing Mutations*. On the shift of expertise, see Gaudillière, "Whose Work Shall We Trust?" On prenatal diagnosis, see Rapp, *Testing Women*; Cowan, *Heredity and Hope*; Stern, *Telling Genes*;

Löwy, *Imperfect Pregnancies*. On the link to eugenics, see Kevles, *In the Name of Eugenics*; Cowan, *Heredity and Hope*.

12. "Chromosomes of Man" (editorial), 716.

13. "Mongolism" was first described as a separate case of mental disability in 1866 by the British physician John Langdon Down who interpreted it as an atavism; see Down, "Observations." In 1961, a group of geneticists, including Ford, Lejeune, Penrose, Polani, and Turpin, signed a letter to the *Lancet* calling for the abandonment of the term *mongolism* from the medical vocabulary. With time the term *Down syndrome* prevailed in both the medical and the public discourse; see Allen et al., "Mongolism"; Rodríguez-Hernández and Montoya, "Fifty Years of Evolution." I follow this later terminology, except in cases where I directly refer to historical texts.

14. L. S. Penrose to J. B. S. Haldane, 4 June 1959, file 136, Penrose Papers, UCL Library Special Collections. On the sequence of events leading to the description of the first chromosome anomalies, see also Kevles, *In the Name of Eugenics*, 238–50.

15. Kevles, *In the Name of Eugenics*, 208.

16. Davenport, "Mendelism in Man" (also quoted in Harper, *First Years*, 57). On the experiment and the ethical questions it raises, see Lombardo, "Tracking Chromosomes."

17. Penrose, *Clinical and Genetic Study*; Penrose, "Maternal Age," 1149–50. On Penrose's early studies of Down syndrome in the context of his broader critical study of the causes of mental disability in a Colchester mental hospital, see Kevles, *In the Name of Eugenics*, 148–63. The striking feature of the Colchester study is the many tables that accompany the report. It was with careful numerical and statistical analyses based on extensive clinical and familial studies that Penrose aimed to counter misleading, simplistic generalizations of the complex etiology of mental disorders. On the pervading eugenic legacy of the interest in the inheritance of mental diseases, see Mazumdar, *Eugenics, Human Genetics and Human Failings*.

18. Mittwoch, "Chromosome Complement."

19. "Mongolism," talk presented at "Discussion on Human Chromosomes in Relation to Disease in Childhood," R. S. M. Paediatrics Section, 27 May 1960 (typescript), p. 1, file 62/5, Penrose Papers, UCL Library Special Collections. Curiously, in a historical review of human chromosome research, Penrose also remarked that, before 1932, "Waardenburg's (1932) belief that mongolism was a monosomic or trisomic condition were not taken seriously." Similarly, "the idea that any person could survive with [an] aberrant karyotype in all cells was . . . rejected because it put the human race on a par with the vegetable kingdom"; Penrose, "Human Chromosomes," 314–15. The suggestion by Penrose and others of a chromosome anomaly as cause of Down syndrome before 1959 is care-

fully reviewed in Carter, "Early Conjectures," even if the singular focus on non-disjunction—the mechanism eventually accepted as basis for the observed trisomy in individuals with Down—somewhat skews the analysis.

20. Turpin, Caratzali, and Rogier, "Étude étiologique," 158. On this point, see also Turpin and Caratzali, "Remarques"; Laplane, "Éloge"; Couturier-Turpin, "La découverte." On the "natalist" context of Turpin's work and its legacies in the Centre de progénèse, founded by Turpin and later directed by Lejeune, see Gaudillière, "Le syndrome nataliste" and "Whose Work Shall We Trust?"

21. For a somewhat different reconstruction of events in Paris, mostly based on an interview with Lejeune, see Kevles, *In the Name of Eugenics*, 245–47. See also note 50 in this chapter.

22. The Paris group worked with connective tissue. This, too, required operative removal.

23. Gautier and Harper, "Fiftieth Anniversary of Trisomy 21."

24. Harper, *First Years*, 63.

25. Lejeune, Gautier, and Turpin, "Les chromosomes humains"; Lejeune, Gautier, and Turpin, "Étude des chromosomes."

26. Lejeune, Turpin, and Gautier, "Le mongolisme." The French group's declaration of Down syndrome as a "chromosome disease" followed a decade after Linus Pauling and his coauthors, with similar emphasis, presented sickle cell anemia as the first "molecular disease"; see Pauling et al., "Sickle Cell Anemia, a Molecular Disease."

27. Interview with Patricia Jacobs by Peter Harper, Salisbury, 13 February 2004, GenMedHist.

28. Apparently in this case the initiative to look at the chromosome configuration in Down syndrome came from Ford rather than from Penrose. Ford made the suggestion following his involvement with the work on Turner and Klinefelter syndromes and drawing on observations of trisomies in mice. Two days after he heard about Lejeune's suggestion, he confirmed the results. See interview with Charles Ford by Daniel Kevles, Abbington, 25 June 1982, Kevles Papers, FA497, box 1, folder 11 (Ford I, side 2, 21–22), RAC.

29. Polani et al., "Mongol Girl." On the visual practices leading to the interpretation of the case and its role in establishing the clinical potential of chromosome analysis, see Santesmases, "Size and the Centromere."

30. Barr and Bertram, "Morphological Distinction."

31. Moore and Barr, "Smears." The cat experiments, performed at the Institute of Aviation Medicine in London, Ontario, were meant to simulate the effect of fatigue in pilots—a phenomenon Barr had experienced firsthand during his wartime service as a pilot in the Canadian Air Force. On the complex process that led to the association of the sex chromatin with the sex chromosomes and the impact of the test on the burgeoning field of sex research, see F. Miller,

"'Your True and Proper Gender'"; Richardson, *Sex Itself*; Ha, "Diagnosing Sex Chromatin." For more on the Barr body test and its impact on chromosome research, see chapter 3 in this volume.

32. On Polani's role in investigating the chromosomal basis of Turner and Klinefelter syndromes, see his unpublished autobiographical essay "The year is 1959 . . ." (2003), Paul Polani, Personal Information file, Royal Society Library and Archives; and interview with Paul Polani by Peter Harper, London, 12 November 2003, GenMedHist. See also Kevles, *In the Name of Eugenics*, 242–45; Harper, "Paul Polani"; Ha, "Marking Bodies," chap. 4; Ha, "Diagnosing Sex Chromatin."

33. Polani, Lessop, and Bishop, "Colour-Blindness," 119–20.

34. Paul Polani, "The year is 1959 . . ." (2003), p. 1; and Harper, "Paul Polani," 726. For more on the role of the X and Y chromosomes as constructed through the studies of sex chromosome anomalies, see chapter 3 in this volume.

35. Polani, Lessop, and Bishop, "Colour-Blindness," 119.

36. Polani, "The year is 1959 . . ." (2003), p. 3. More generally, the Symposium on Nuclear Sex, convened in September 1957 by the clinical pathologists David Robertson Smith and William M. Davidson of King's College Hospital Medical School in London, proved an important occasion for bringing together clinicians, cytologists, and human cytogeneticists. The meeting was prompted by Murray Barr's visit to the United Kingdom; see "Autobiographical Notes," vol. 1, p. 40, Murray L. Barr Papers, National Archives, Ottawa. It was followed by a second meeting, two years later, dedicated to human chromosomal abnormalities. This meeting responded to the technological shift under way from chromatin testing to chromosome analysis; see Robertson Smith and Davidson, *Symposium*; Davidson and Robertson Smith, *Proceedings*.

37. C. E. Ford et al., "Sex-Chromosome Anomaly," 712. Barr, the inventor and keen promoter of the test that carried his name, had made a similar cautioning suggestion a few years earlier; see Barr, "Cytological Tests"; Ha, "Diagnosing Sex Chromatin," 72. Barr himself had tried to interest the cytogeneticist Hsu, among others, to help him sort out the chromosomal basis of his test. Hsu was receptive but did not get fully engaged in the issue; Ha, "Diagnosing Sex Chromatin," 73–74. For the recollection of one of the coauthors of the article on Turner syndrome, the Brazilian endocrinologist José Carlos Cabral de Almeida, then visiting Polani's laboratory, on the team effort leading to the publication, see Jung et al., "Revisiting Establishment."

38. For the following reconstruction of events, especially regarding Polani's role, see Ha, "Marking Bodies," chap. 4, and Harper, "Paul Polani," in addition to the relevant scientific articles.

39. Ford, Jacobs, and Lajtha, "Human Somatic Chromosomes."

40. In a later interview Jacobs stamped the quality of the preparations as straightforwardly "terrible" and a "disaster." The trouble started when Jacobs,

then still "very young and very naïve," was dispatched to King's College Hospital, London, to collect the fresh sternal puncture specimen. Rather than being led straight into the operating room, where the patient was ready for the intervention, she was made to wait in the waiting room as a regular outpatient. When the mistake was finally discovered, the patient had been waiting for an hour and everyone was "furious"; interview with Patricia Jacobs, Salisbury (UK), 2 June 2010.

41. Interview with Charles Ford by Daniel Kevles, Abbington, 25 June 1982, Kevles Papers, FA497, box 1, folder 11 (Ford I, side 2, 25), RAC.

42. Barr, "Human Cytogenetics," 82. Various interviews confirmed such complaints. On the publication policy of the *Lancet* with respect to chromosomes, see also Harper, *First Years*, 98–99. It should be pointed out that, before the many case reports about chromosome anomalies, studies on "nuclear sexing" were also very well represented in the *Lancet*.

43. For the British context, see Sturdy and Cooter, "Science, Scientific Management, and the Transformation of Medicine"; and Coventry and Pickstone, "From What and Why Did Genetics Emerge."

44. Lejeune, "Le mongolisme."

45. Rostand, "Parviendra-t-on bientot a guérir les enfants?"

46. See, for example, Fraccaro and Lindsten, "Le malattie cromosomiche"; and Buzzati-Traverso, "Tutti da rifare"; Buzzati-Traverso, "La materia e la vita." Although Buzzati-Traverso is best remembered for his effort to launch an international molecular biology institute in Naples in the 1960s, he was an enthusiast of chromosome research. Preserved in Fraccaro's archive is a card, signed by Buzzati-Traverso, sporting a photograph of his chromosomes and the message: "My chromosomes shall bring you my fond wishes for a prosperous 1961"; Fraccaro Collection. On Buzzati-Traverso and the rise of molecular biology in Italy, see Capocci and Corbellini, "Adriano Buzzati-Traverso."

47. Interview with Marco Fraccaro, Pavia, 30 March 2004.

48. Redding and Hirschhorn, *Guide*, 1.

49. Paul and Brosco, *PKU Paradox*.

50. The respective contributions of Turpin, Lejeune, and Gautier in the 1959 work leading to the first description of trisomy 21 has given rise to a somewhat bitter controversy among the participants and their descendants. In a recent memoir Gautier (whose name was misspelled in the key publication) expressed the extent to which she felt marginalized in the group effort that led to the discovery and her resentment about Lejeune taking most of the credit. According to Gautier, Turpin also came to resent Lejeune's behavior; Gautier and Harper, "Fiftieth Anniversary of Trisomy 21." Turpin and Lejeune collaborated closely up to the mid-1960s, when Turpin retired. At a certain point, tensions developed, although the exact reasons for this are somewhat unclear. Various writings, collected in a biographical file at the Pasteur Institute deposited by Turpin's fam-

ily, including Turpin's eloge by the French National Academy of Medicine and a short popular history of the events leading to the discovery of trisomy 21 by Turpin's daughter, herself an accomplished cytogeneticist, stressed Turpin's decisive contribution to the discovery; see folder Documents à charactère biographique, Fond d'archives Raymond Turpin, FR AIR TRP, Service des archives, Institute Pasteur, Paris. In the 1970s, Lejeune, then the director of the cytogenetics laboratory at the Hôpital Necker—Enfants Malades, took a vocal antiabortionist stance, distancing himself from many geneticists. This development, together with tensions between Tjio and Levan regarding their respective contribution to the establishment of the new number of human chromosomes, might have cost the field of cytogenetics its Nobel Prize, which supposedly Levan, Lejeune, and Caspersson were going to share; interview with Bernhard Dutrillaux, Paris, 30 June 2011.

51. On the expert panel, see pp. 92–96 in this volume. On Lejeune's role as a French scientific envoy, see interview with Carlos Sonnenschein, São Paulo, Brazil, 20 July 2017. In 1961, Sonnenschein participated in a course on chromosome techniques offered by Lejeune in Buenos Aires that set him on a new research career in cytology. Unfortunately, Lejeune's personal archive, though apparently extant, has not yet been made publicly available.

52. This was not an isolated attempt. Lejeune also strived to treat fragile X syndrome, another newly described chromosome disease leading to mental disability, by administrating patients high doses of folic acid; see Hogan, "Disrupting Genetic Dogma," 184–85. On the efforts to treat genetic diseases through nutritional and metabolic interventions in the 1960s and 1970s, see also Löwy, *Imperfect Pregnancies*, 64–68.

53. Sex chromatin analysis of amniotic fluid with the aim of predicting the sex of the fetus became possible in the mid-1950s. It found limited clinical application in cases of sex-linked diseases like muscular dystrophy or hemophilia that run in families; see Cowan, "Aspects." The ability to grow amniotic fluid cells in culture and perform a chromosome test was first reported in the mid-1960s; Steele and Breg, "Chromosome Analysis." Prenatal biochemical tests for neural tube defects (or spina bifida) and other conditions became available in the following years.

54. For his stand against abortion and his care of Down syndrome children, Lejeune's heirs petitioned for his beatification. The case remains open. As argued in detail by Gaudillière, karyotyping followed a different trajectory in the genetic service run by Maurice Lamy at the same hospital; Gaudillière, "Whose Work Shall We Trust?" The technique was introduced in Lamy's department in 1959 by a student, Jean de Grouchy, who had spent time in Penrose's laboratory. De Grouchy employed chromosome analysis to revise the clinical classification and causal description of certain diseases and to drive clinical innovation. The counseling service was used to collect data on transmission patterns

and incidences of genetic diseases as well as to advise on risk factors. The service grew into a center for medical cytogenetics in France. De Grouchy's work culminated in the publication of the *Clinical Atlas of Human Chromosomes*, a compilation and visual guide to all known chromosome syndromes that paired cytogenetic data with detailed clinical descriptions and case photographs; de Grouchy and Turleau, *Clinical Atlas*. Despite their different professional trajectories Lejeune and de Grouchy reportedly were close personal friends; see Harper, *First Years*, 157.

55. The most comparable case in the British context is that of Ferguson-Smith in Glasgow; see pp. 62–63 in this volume.

56. Latour, "Science in Action."

57. Interview with Charles Ford by Daniel Kevles, Oxford, 25 June 1982, Kevles Papers, FA497, box 1, folder 11, RAC. Unfortunately, Ford's papers have not been preserved.

58. On Ford's move away from human cytogenetics for both personal and institutional reasons, see interview with Edward (Ted) Evans by Peter Harper, 23 August 2004, GenMedHist.

59. The original endowment provided by the National Spastics Society was £2 million; see [Paul Polani], "The Paediatric Research Unit (a Brief Outline)" (typescript, undated, updated 2003), Polani, Personal Information file, Royal Society Library and Archives. The existing online history of the Spastics Society (now Scope) hardly mentions the grant to Polani's unit except in a brief interview, where it becomes clear that the decision was hard won to put the money into research that was not exclusively dedicated to cerebral palsy but to the causes of developmental disability more generally; Davies, *Changing Society*, 153. Unfortunately, no archival records seem to have survived either at Guy's Hospital or at Scope that would make it possible to investigate the setting up of the unit in any further detail. Also Polani's original blueprint for the unit does not seem retrievable. I thank Elizabeth Manners at Guy's Hospital for guidance and access to the internal records of the Paediatric Research Unit kept at what is now the Department of Medical and Molecular Genetics at King's College London School of Medicine. For the breadth of research pursued at the Paediatric Research Unit, see the Festschrift for Polani: Adinolfi et al., *Paediatric Research*.

60. The genetic counseling clinic started by Fraser Roberts at Great Ormond Street in 1946 is often considered the first such service in Europe and the third worldwide, after the Hereditary Clinic at the University of Michigan and the Dight Institute at the University of Minnesota. Regarding Fraser Roberts's active membership in the Eugenic Society from 1926 to 1971, Polani remarked that he considered him a "mild reform eugenicist" who, in his later years, "rather forgot eugenics and thought more of individual consultancies and patients"; Polani, "John Alexander Fraser Roberts," 321.

61. On the development of the fetus as an experimental object, see Hopwood, "Embryology"; Lynn Morgan, *Icons of Life*; Maienschein, *Embryos under the Microscope*.

62. P. Polani, "The year is 1959 . . ." (2003, typescript), Paul Polani, Personal Information file, Royal Society Library and Archives; and interview with Paul Polani by Peter Harper, Guildford, 12 November 2003, GenMedHist.

63. Polani, "Incidence of Chromosomal Malformations."

64. Jacobs, "William Allan Memorial Award Address," 695.

65. Benirschke et al., "Standardization."

66. According to the reminiscences of a young geneticist sitting in the audience, when German-born biologist Leo Sachs from the Weizmann Institute in Israel—in a lecture presented at the first International Conference of Human Genetics in Copenhagen in 1956—suggested that sex could be analyzed prenatally through amniocentesis using the Barr body test, "nobody was shocked"; see Sachs et al., "Prenatal Diagnosis"; and interview with Marco Fraccaro, 30 March 2004, Pavia. Other groups were pursuing similar experiments; see especially Fuchs and Riis, "Antenatal Sex Determination"; Fuchs et al., "Antenatal Detection." However, it took a decade until amniocentesis could be safely applied later in the pregnancy and fetal cells from the amniotic fluid cultured for chromosome analysis; see Steele and Breg, "Chromosome Analysis." For a multifaceted account of the biomedical, legal, political, and sociocultural developments that led to the establishment of amniocentesis and prenatal diagnosis as an accepted practice, see Rapp, *Testing Women*, esp. 23–52. On prenatal diagnosis, see also Löwy, *Imperfect Pregnancies*.

67. Ferguson-Smith's cytogenetic service in Glasgow started offering prenatal diagnosis at about the same time; see Ferguson-Smith, "Putting Medical Genetics into Practice" and later in this section. They both claim priority.

68. Polani et al., "Sixteen Years' Experience."

69. Interview with Paul Polani by Peter Harper, London, 12 November 2003, GenMedHist.

70. See "Paediatric Research Unit—Guy's Hospital Medical School: An Abridged Record of Research and Service" (typescript, presented by Polani to the Royal Society on 7 September 1981), Polani, Personal Information file, Royal Society Library and Archives.

71. As has been argued for the case of Manchester, the regional genetic services exemplified a new linkage between research projects, emerging medical specialists, hospital services, National Health Service policies to introduce cost-effective interventions, and patient service demand; Coventry and Pickstone, "From What and Why Did Genetics Emerge." On the setting up of the regional genetic centers, see also Leeming, "Tracing the Shifting Sands."

72. For a population based study, see J. R. Fraser Roberts and P. E. Polani, "South-East Regional Survey: A Collaborative Scheme for Studying Certain Im-

portant Chromosomal and Other Congenital Defects, and Determining Their Frequency in the Population (Explanatory Notes)" (1966), Polani, Archival Collection, Guy's Hospital, King's College London School of Medicine, Department of Medical and Molecular Genetics.

73. Santesmases, "Human Autonomous Karyotype."

74. Polani et al., "Sixteen Years' Experience," 170.

75. Stern, *Telling Genes*, 47–48, 154. In her detailed ethnography of prenatal diagnosis, Rayna Rapp remarked that, in general, women reacted much more strongly to the ultrasound images rather than to the "caterpillar stick figures" of chromosome photographs; Rapp, *Testing Women*, 221. Yet this does not preclude physicians and counselors from making strategic use of the latter in the consultation room as observed by Stern. On the rhetorical strength of visual evidence in genetic diagnostics, see also Taussig, *Ordinary Genomes*, 111–26. As an artist in residence at Guy's Hospital Genetics Department in 2002 and at the hospital's Cytogenetics Assisted Conception Unit in 2008, Gina Glover created several artworks based on chromosome images produced in the laboratory that visitors and patients in these units can see hanging on the walls. The artworks include the prize-winning *Chromosome Socks*, composed of twenty-four ordered pairs of striped socks, apparently collected from cytogeneticists in the lab, and *Painting with Light: Mosaic of Images and Quotations from the Cytogenetics Department, Guy's Hospital*, incorporating Polani's microscope filters; see Glover's website at http://www.ginaglover.com, as well as the website Art in Hospitals, at http://www.artinhospitals.com/photo_genetics_chromosomes.html. Glover's presence in the laboratory and her artwork point to the continuing impact of chromosomal imagery in the clinical context.

76. Survey-based studies were not absent in Polani's unit but generally they built on data won through the services offered and they did not necessarily involve just cytogenetic data. On Court Brown's earlier career, see pp. 22–24 in this volume.

77. "Proposed M.R.C. Group for Research on the General Effects of Radiation at the Western General Hospital, Edinburgh. Draft" (undated but probably 1955); MRC Human Genetics Unit, Library and Archive, Edinburgh.

78. "Notes on meeting in Dr. Cowan's office at Edinburgh on the 6th January" (1956), p. 2; MRC Human Genetics Unit, Library and Archive, Edinburgh.

79. [W. M. Court Brown], "Memorandum on a Proposed Registry of Persons with an Abnormal Chromosome Constitution" (draft, 1959), file FD 12/445, NA. Registries were established tools for epidemiological research. According to a definition offered by WHO, a registry (in contrast to a simple notification) requires a permanent record, that the cases are followed up, and that statistical tabulations are prepared; World Health Organization Expert Committee on Health Statistics, *Epidemiological Methods*, 11. Supposedly the oldest registry in this sense was a leprosy registry, established in Norway in 1856; Irgens, "Ori-

gin." Registries gained increasing importance in the course of the twentieth century, especially with respect to the epidemiological study of cancer and mental diseases.

80. "Group for Research on the General Effects of Radiation, Proposed Registry of Abnormal Human Karyotypes," January 1960, file FD 23/142, NA. See also "Notes on Procedure: Types of Case [*sic*] to Be Registered"; ibid.

81. See correspondence in file FD 12/445, NA.

82. D.R. (initials) to B. L. Lush, 6 October 1959 (MRC internal note on commentary card accompanying folder), file FD 23/220, NA.

83. The first neonatal screening program based on buccal smears was started by Keith Moore, Barr's student, in London, Ontario, but the Edinburgh study soon grew into the largest of its kind.

84. For further details of the prospective studies, see pp. 98–105 in this volume.

85. Interview with Patricia Jacobs, Salisbury, 2 June 2010.

86. W. M. Court Brown, "Memorandum on Current Demands for Cytogenetic Studies and the Requirements for a Routine Service," 7 November 1962 (typescript), file FD 23/148, NA.

87. W. M. Court Brown, "The Provision of Facilities in Human Cytogenetics to Meet the Needs of Medical Practice," 21 September 1966, file FD 9/1281, NA. On another occasion Court Brown suggested that the most satisfactory way of organizing a service would be to set up of a few "expert centres," each capable of covering a population of about five million people; see "Extract from CRB [Cell Research Board], Minutes of Meeting of March 1967, item 38. Noon session: Developments in the Clinical Aspects of Cytogenetics. Talk by Dr. Court Brown," MRC 67/384, file FD 9/1281, NA. The need for expert supervision of cytogenetic work was demonstrated by the fate of the cytogenetic program at the Psychiatric Genetics Research Unit at the Maudsley Hospital in London. A cytogenetic laboratory was established at the unit in 1963, but the work was considered of such poor quality that funding was discontinued a few years later; see file Clinical Research Board—Visit to Psychiatric Genetics Research Unit (E. T. O. Slater), FD 9/1295, NA.

88. [W. M. Court Brown], "Developments in the Clinical Aspects of Cytogenetics" (undated, typescript attached to letter of 19 October 1966), file FD 9/1281, NA. Autoradiography was a technique based on radioactive labeling that helped identify single chromosomes. It was soon superseded by other banding techniques. It regained importance as an approach for in situ hybridization before fluorescent labeling again rendered it obsolete; see chapter 5 in this volume.

89. W. M. Court Brown, "The Provision of Facilities in Human Cytogenetics to Meet the Needs of Medical Practice," 21 September 1966, file FD 9/1281; W. M. Court Brown, "Contributions of Human Cytogenetics to Clinical Medicine," 16 March 1967, file FD 9/1281, NA.

90. Comfort, *Science of Human Perfection*.

91. The project drew on a considerable number of Klinefelter patients whom Ferguson-Smith had identified performing buccal smear screens of men attending the infertility clinic and at a local hospital for the disabled. The project was initiated by the pathologist Bernard Lennox, who had worked with Polani on Turner syndrome and meanwhile had moved from London to Glasgow. In both cases the discrepancies between anatomical and nuclear sex prompted the effort to look at the chromosomes of the patients at a time when the techniques were only just becoming available. For more on the Barr body test, the implications for individuals with Turner or Klinefelter syndrome and the incipient sex chromosome analysis to adjudicate the cases, see chapter 3 in this volume.

92. Ferguson-Smith, "Cytogenetics," 7. Like Ferguson-Smith several other researchers moving into cytogenetics in the 1950s had a background in pathology, bringing with them their skill of working with the microscope.

93. Ferguson-Smith, "Putting Medical Genetics into Practice." On Ferguson-Smith's career, see also interview with Malcolm Ferguson-Smith by Peter Harper, Cambridge, 5 December 2003, GenMedHist; interview with Malcolm Ferguson-Smith, Cambridge, 28 January 2013; and interview with Marie Ferguson-Smith, Cambridge, 5 July 2014.

94. V. McKusick to W. M. Court Brown, 31 May 1961, and W. M. Court Brown to V. McKusick, 5 June 1961, file Ferguson-Smith, Malcolm A., box 509 284, Alan Mason Chesney Medical Archives, Johns Hopkins Medical Institutions (Hopkins Medical Archives). For the problem in recruiting someone to succeed Ferguson-Smith, see further correspondence in the same box.

95. Interview with Barbara Migeon by Jennifer Caron and Nathaniel Comfort, Johns Hopkins University Medical School, Baltimore, 2 June 2005, OHHGP. Also, interview (by author) with Barbara Migeon, Baltimore, 1 May 2014.

96. Comfort, *Science of Human Perfection*, 183.

97. See chapter 5 in this volume.

98. Canguilhem, *On the Normal and the Pathological*.

99. Hopwood, "Visual Standards."

100. On the standardization of the human karyotype, see also Lindee, *Moments of Truth*, 90–119. More generally, on standards and standardization in science, see Wise, *Values of Precision*; Bowker and Star, *Sorting Things Out*; Timmermans and Epstein, "World of Standards." On the history and normativity of the "normal" and the rise to power of the concept in the mid-twentieth century, see Cryle and Stephens, *Normality*.

101. Painter, "Studies in Mammalian Spermatogenesis II," plate 6; Hsu, "Mammalian Chromosomes in Vitro," fig. 14.

102. Tjio and Levan, "Chromosome Number of Man," 3.

103. Albert Levan, "Conference on Human Chromosomes—Colorado,

April 8–11, 1960. Comments on points A-G of the provisional agenda" (typescript, undated), p. 6, file Normal Karyotype, Fraccaro Collection. On the transition from drawing to photographing chromosomes and the impact on the practice of cytogeneticists, see also de Chadarevian, "Chromosome Photography."

104. Albert Levan, "Conference on Human Chromosomes—Colorado, April 8–11, 1960. Comments on points A-G of the provisional agenda" (typescript, undated), pp. 3–4, including tables and figures, file Normal Karyotype, Fraccaro Collection.

105. Human Chromosomes Study Group, "Proposed Standard of Nomenclature," 1.

106. Puck, "Living History Biography," 281.

107. Puck, "Living History Biography," 281.

108. Human Chromosomes Study Group, "Proposed Standard of Nomenclature," 1.

109. Human Chromosomes Study Group, "Proposed Standard of Nomenclature," 6, table 1.

110. Hsu, *Human and Mammalian Cytogenetics*, 59.

111. Robinson, "Living History," 478.

112. Human Chromosomes Study Group, "Proposed Standard of Nomenclature," 3.

113. Clarifying an existing confusion in the literature, the study group recommended using the term *karyotype* to define an ordered arrangement of the chromosomes of a single cell, either drawn or photographed, while the term *idiogram* was to be reserved for the diagrammatic representation of a karyotype; Human Chromosomes Study Group, "Proposed Standard of Nomenclature," 1n.

114. Human Chromosomes Study Group, "Proposed Standard of Nomenclature," 3.

115. Penrose, "Proposed Standard System."

116. Penrose, "Proposed Standard System," fig. 1, caption.

117. See, e.g., Penrose, "London Conference"; Penrose, "Note on the Measurements"; Penrose, "Introductory Address"; Penrose, "Human Chromosomes." Copious material, including notebooks, tables, graphs, and correspondence, in Penrose's papers dating from the early to mid-1960s further confirms the importance of the issue for Penrose; see Penrose Papers, files 90/1A-C and 90/2, UCL Library Special Collections. On the importance of quantification for Penrose, see also Kevles, *In the Name of Eugenics*, 221–22. Not everyone agreed with Penrose's insistence on measurement. David H. Carr, a collaborator in Barr's laboratory at the University of Western Ontario and one of Penrose's correspondents on the matter, contended: "I am enclosing measurements which you requested. Since the early days of my work we have never measured chromosomes as we did not feel it contributed to the identification or understanding of

human karyotypes. These measurements were made specifically and with all due care. The results (unless my eye or arithmetic is grossly out of line) support the view that measurement does not assist in pairing or identifying chromosomes." D. H. Carr to L. Penrose, 14 August 1963, file 90/2, UCL Library Special Collections. Supporting Penrose's position on the matter, Marie Ferguson-Smith, who did much of the cytogenetic lab work in the unit set up by Malcolm Ferguson-Smith at Johns Hopkins in Baltimore and later in Glasgow, excitedly wrote to McKusick about a new method she had devised that produced "much better and accurate measurements." It involved projecting the photographic film onto a white sheet of paper and tracing the chromosomes on the image. She added: "I don't know why we did not think of that before. The whole process is so much quicker." Marie Ferguson-Smith to V. McKusick, 15 June 1962, Malcolm A. Ferguson-Smith, Biographical File, Hopkins Medical Archives.

118. See file Normal Karyotype, Fraccaro Collection. The letters later assigned to each chromosome group were added in pencil.

119. Human Chromosomes Study Group, "Proposed Standard of Nomenclature." See frontispiece in this volume.

120. Interview with Marco Fraccaro, Pavia, April 2004. According to Fraccaro, the figures included in the article were probably supplied by Hamerton, Polani's collaborator at the Paediatric Research Unit at Guy's Hospital.

121. Hsu, *Human and Mammalian Cytogenetics*, 61.

122. This can also be seen as the gist of Lejeune's continuing misgivings. In the context of a discussion on projects in human genetics that the WHO could usefully support, Lejeune explained his stance thus: "The problem of human chromosome classification is still open and effectively the gentleman agreement reached in Denver is not the best solution. My very criticism against it is that the numerical classification is much more precise than our present standards of recognition. But how could WHO sponsor a regressive nomenclature? In my opinion the time is not yet ripe for a useful revision of the actual numbers"; J. Lejeune to R. Lowry Dobson, 5 January [June?] 1962, file G3-370-2, fiche 1; Plan for a Medical Research Program Related to Human Genetics (1959–1965), WHO Archives. Other researchers shared Lejeune's concern that current techniques did not support the numbering of single chromosomes (especially in group 6–12) as required by the Denver nomenclature. The point was made most forcefully by Klaus Patau, a German émigré at the University of Wisconsin; Patau, "Identification of Individual Chromosomes"; Patau, "Chromosome Identification." Although already working on human chromosomes, he had not yet published a human karyotype at the time of the Denver conference and therefore was not invited to the meeting. A few months later he was the leading author on two papers that reported the first case of trisomy 13 (also known as Patau syndrome) and independently described a case of trisomy 18, although the first

observation of the syndrome is attributed to John Edwards, then at Oxford. On Patau and the problem of stabilizing the description of chromosomal syndromes, see Löwy, *Imperfect Pregnancies*, 49–53.

123. Interview with Marco Fraccaro, Pavia, April 2004.

124. *46—The Human Chromosome Newsletter*, no. 1, October 1960, and *46—The Human Chromosome Newsletter*, no. 9 (1963), editorial note, Fraccaro Collection. The newsletter *Mammalian Chromosomes* compiled by Hsu in Houston was started a few months earlier (May 1960). In a later issue, Hsu addressed the question concerning the division of labor with the *Human Chromosome Newsletter* thus: "I have not corresponded with Dr Harnden and Miss Jacobs concerning this matter, but tentatively we shall propose that the *Mammalian Chromosomes Newsletter* publish raw data on human cytogenetics as we have been practicing. Articles relating to negative data, advanced abstracts of findings on human chromosomes and their anomalies, and other subjects on human karyology should be submitted to the *Human Chromosome Newsletter*"; *Mammalian Chromosomes Newsletter*, no. 9, January 1963, p. 2, Fraccaro Collection. Following consultation at the next standardization meeting in Chicago, publication of the *Human Chromosome Newsletter* ceased in 1966 after twenty issues. It was believed that, while at the start the newsletter had served "a worthwhile scientific purpose," this had become less so as other journals and improved abstract services were providing similar functions; P. A. Jacobs and D. G. Harnden (editorial), *46—The Human Chromosome Newsletter*, no. 20, December 1966, p. 1. On the role of newsletters in the moral economy of science and as historiographical tools, see Kelty, "This Is Not an Article."

125. Bangham, "Writing, Printing, Speaking."

126. Hamerton et al., "London Conference."

127. Bergsma, Hamerton, and Klinger, "Chicago Conference," 1.

128. Robinson, "Living History," 478. Other interpretations offered are that q was a typo for the letter g (*grand* in French) or that q stemmed from the formula $p + q = 1$, according to which the short and long arm together form the whole chromosome; see Harper, *First Years*, 148; Hogan, *Life Histories*, 60–61.

129. Bergsma, Hamerton, and Klinger, "Chicago Conference," 6.

130. Bergsma, Hamerton, and Klinger, "Chicago Conference," 7.

131. Hamerton, Jacobs, and Klinger, "Paris Conference"; Hamerton and Klinger, *Paris Conference (1971)*. See also figure 5.4 on p. 163 in this volume.

132. Hamerton and Klinger, *Paris Conference (1971)*, 14–28.

133. On visual standards, see Hopwood, "Visual Standards." An influential example for annotated chromosome diagrams can be found in Victor McKusick's effort to map the genes responsible for human diseases; McKusick, *Mendelian Inheritance in Man*. On chromosome diagrams as representational devices and their function as "infrastructure," see also Hogan, "'Morbid Anatomy'"; Hogan, *Life Histories*; and chapter 5 in this volume.

134. Bergsma, *ISCN 1978*.

135. On the preservation of individual "styles" of cytogenetic laboratories in preparing and presenting chromosomes up to this day, despite the widespread standardization of tools, see Turrini, "Continuous Grey Scales."

136. Ford, Jacobs, and Lajtha, "Human Somatic Chromosomes." Of special interest to human geneticists was the comparison to other primates and especially anthropoid apes. For an example of one such study performed by an international group of researchers on two gorillas from the local zoo on the sidelines of a WHO international laboratory course for cytogenetic practices in Basel, see Hamerton et al., "Somatic Chromosomes of the Gorilla." The researchers used the new skin culture technique that made such a study possible.

137. Lindee, *Moments of Truth*, 1; McKusick, "Human Genome," 7; Comfort, *Science of Human Perfection*, 198.

138. On this point, see also Coventry and Pickstone, "From What and Why Did Genetics Emerge."

139. Polani, "Human and Clinical Cytogenetics," 119.

140. "Mongolism," talk presented at "Discussion on Human Chromosomes in Relation to Disease in Childhood," R. S. M. Paediatrics Section, 27 May 1960 (typescript), p. 1, file 62/5, Penrose Papers, UCL Library Special Collections. Interestingly, Court Brown, who had an expansive vision of the potential of human karyotyping, cautioned that pediatricians had perhaps been "overenthusiastic about the value of cytogenetics in mongolism" and that "here its value has become exaggerated and perhaps distorted." In Court Brown's view, "careful clinical appraisal still remains the best and the most economic method of diagnosis, with cytogenetics being employed only to sort out the small hard core cases where there is some doubt on clinical grounds"; W. M. Court Brown, "Contributions of Human Cytogenetics to Clinical Medicine," MRC 67/357–CR 67/26, 16 March 1967, p. 2, file FD 9/1281, NA.

141. Polani, "Human and Clinical Cytogenetics," 119.

Chapter Three

1. De Grouchy and Turleau, *Clinical Atlas*, ix. Writing in the mid-1970s, the authors of the *Clinical Atlas of Human Chromosomes*, for which they originally considered the title "From 1 to 22, and X and Y," described the "personality" of each chromosome as made up of the characteristic banding pattern it displays, its evolutionary history, the number and kind of genes it carries, and its known disease markers; de Grouchy and Turleau, *Clinical Atlas*, ix.

2. Mittwoch, *Sex Chromosomes*; Richardson, *Sex Itself*.

3. The poem—probably just one of many more that exist somewhere—is reproduced in Sarah Richardson's exploration of the impact of gender stereotypes

in the history of sex chromosome research. Composed by an anonymous author, it was included on a loose sheet of paper in a privately owned copy of a genetic textbook by Thomas Hunt Morgan and his colleagues and eventually made it into a scanned copy of the book that is now publicly available; see Richardson, *Sex Itself*, 61–62.

4. Hamerton, *Human Cytogenetics*.

5. Richardson, *Sex Itself*.

6. On the role of controversies in the formation of scientific fields, exemplified for the case of behavioral genetics, see Panofsky, *Misbehaving Science*.

7. On the independent and equally important contributions of both Stevens and Wilson to the chromosomal theory of sex, see Richardson, *Sex Itself*, 29–34.

8. Mittwoch, *Sex Chromosomes*, 218; Carlson, *Mendel's Legacy*, 79–98.

9. Painter, "Studies in Mammalian Spermatogenesis II"; Painter, "Sex Chromosomes of Man." On Painter's research career see the biographical memoir by his student Bentley Glass; Glass, "Theophilus Shickel Painter."

10. According to his biographer, Painter described the thick chromosome strands in the nuclei of the salivary glands "independently but simultaneously with E. Heitz and Hans Bauer in Switzerland"; see Glass, "Theophilus Shickel Painter," 321.

11. Mittwoch, *Sex Chromosomes*, 104.

12. K. L. Moore, introduction to *The Sex Chromatin*, 3.

13. MacLean, "Sex Chromatin Surveys"; Ferguson-Smith, "Sex Chromatin."

14. Court Brown, "Study of Human Sex Chromosome Abnormalities," 392.

15. F. Miller, "'Your True and Proper Gender'"; Ha, "Marking Bodies," chap. 4; Ha, "Diagnosing Sex Chromatin"; Richardson, *Sex Itself*.

16. Ha, "Riddle of Sex."

17. Anomaly to M. Barr, 29 May 1952; M. Barr to Anomaly, 2 June 1952; and M. Barr to Anomaly, 6 October 1952, Murray Barr Papers, file 5-21, National Archives, Ottawa. The letter is cited and effectively used by Ha, who unearthed it from Barr's voluminous correspondence; see Ha, "Marking Bodies," chap. 4; Ha, "Diagnosing Sex Chromatin."

18. For a brief history of the syndrome from a medical genetic point of view, see Ferguson-Smith, "Sex Chromatin."

19. Ford, Jacobs, and Lajtha, "Human Somatic Chromosomes."

20. Jacobs and Strong, "Case of Human Intersexuality."

21. Ferguson-Smith, "Sex Chromatin," 278, table 18-1.

22. Jacobs et al., "Evidence." Anticipating the existence of a "superfemale" karyotype in humans as observed in *Drosophila*, Puck in Denver apparently systematically tested young girls showing early sexual development, hypothesizing that they might carry a triple-X karyotype. This can be gleaned from a popular article written by Buzzati-Traverso, who was visiting Puck's laboratory at

the time; see Buzzati-Traverso, "La materia e la vita." In an earlier article published in the science section of the daily national broad sheet *Il Giorno*, Buzzati-Traverso introduced readers to the role of the Y chromosome in human sex determination; Buzzati-Traverso, "Tutti da rifare."

23. Jacobs et al., "Evidence," fig. 1.

24. Shortly after the identification of Down syndrome as trisomy 21, Ferguson-Smith, together with Levan, who was visiting Ferguson-Smith in Baltimore, established that the anomaly in fact regarded chromosome 22 rather than 21. The two researchers tried to get the numbers corrected. Yet to avoid confusion and to not call into question a recently achieved consensus, the order of the two smallest chromosomes was switched around instead. On the episode, see Ferguson-Smith, "Putting Medical Genetics into Practice," 15; Ferguson-Smith, "Cytogenetics," 6; and interview with Malcolm Ferguson-Smith, Cambridge, 28 January 2013.

25. Bridges, "Sex."

26. "Chromosomes of Man" (editorial), 761.

27. Jacobs, Price, and Law, *Human Population Cytogenetics*, 4.

28. Jacobs and Strong, "Case of Human Intersexuality," 302.

29. Court Brown, "Sex Chromosome Aneuploidy," 33; Ferguson-Smith, "Putting Medical Genetics into Practice," 4.

30. Richardson, *Sex Itself*. According to Richardson, "The story of the gendering of the X and Y as objects of scientific knowledge" took off only in the late 1950s to early 1960s; Richardson, *Sex Itself*, 18. For a focus on the earlier history of sex chromosome research, see Satzinger, *Differenz und Vererbung*.

31. Harnden, "Early Studies," 165.

32. Ohno and Makino, "Single-X Nature"; Ohno, "Single-X Derivation."

33. On the mosaic hypothesis and its clinical impact, see Migeon, *Females Are Mosaics*. Barbara Migeon, a Professor of Pediatrics at Johns Hopkins, dedicated much of her long research career to the development of the X chromosome inactivation and mosaic hypothesis; interview with Barbara Migeon, Baltimore, 1 May 2014.

34. Jacobs et al., "Aggressive Behaviour." On the "masculinizing properties" of the Y chromosome, see Court Brown, "Sex Chromosome Aneuploidy," 33.

35. Interview with Patricia Jacobs, Salisbury, 2 June 2010.

36. Casey et al., "Sex Chromosome Abnormalities."

37. See MacLean et al., "Survey of Sex-Chromosome Abnormalities." The paper provided a table summarizing the data of other available surveys of "mental defectives," categorized according to their IQ into "mild" and "major defectives" and subdivided into males and females. Overall these surveys included 8,000 individuals in addition to the 4,500 covered by the new study; MacLean et al., "Survey of Sex-Chromosome Abnormalities," 293, table 1.

38. Jacobs et al., "Aggressive Behaviour," 1351.

39. Lewontin, Rose, and Kamin, *Not in Our Genes*, 49–52; Richardson, *Sex Itself.*

40. Quoted in Harper, *First Years*, 90. See also interview with Patricia Jacobs, Salisbury, 2 June 2010.

41. Jacobs et al., "Chromosome Studies," 341.

42. Jacobs et al., "Chromosome Studies," 341. The final report listed twenty-seven refusals on 342 patients of both wings; available tests indicated that these patients had higher average IQs than the study participants; Jacobs et al., "Chromosome Studies," 344.

43. Sandberg et al., "XYY Human Male."

44. Hauschka et al., "XYY Man."

45. Jacobs et al., "Aggressive Behaviour," 1352.

46. Harper, *First Years*, 89; Casey et al., "YY Chromosomes."

47. Price et al., "Criminal Patients," 565.

48. "YY Syndrome" (editorial).

49. Park, "YY Syndrome."

50. "Criminal Behaviour" (editorial).

51. Court Brown, "Genetics and Crime," 318.

52. Price and Whatmore, "Criminal Behaviour"; Price and Whatmore, "Behaviour Disorders," 536.

53. Jacobs et al., "Chromosome Studies."

54. Besides the full report of the Carstairs study that appeared shortly before Court Brown's review, the unit had published the results of an extensive chromatin survey including more than twelve thousand inpatients of mental hospitals as well as close to one thousand female outpatients of a psychiatric clinic in Scotland; see Jacobs et al., "Chromosome Studies"; MacLean et al., "A Survey of Sex Chromatin Abnormalities in Mental Hospitals." The study aimed to assess the role of additional X chromosomes in the development of mental illness. A later study, again with Jacobs as leading author, reported on chromosome surveys in a variety of corrective schools and penal institutions through which some of the Carstairs patients had transited. The results showed no increased frequency of males with an XYY chromosome complement; Jacobs et al., "Chromosome Surveys in Penal Institutions." Overall these studies, together with all the other investigations pursued in parallel and described in other chapters, point to the ever-expanding screening efforts of the Edinburgh unit. In 1968, already strongly affected by a first heart attack, Court Brown also published a sixty-page comprehensive review on (XYY and other) sex chromosome anomalies and their relation with mental disability and criminal behavior; Court Brown, "Sex Chromosome Aneuploidy."

55. Court Brown, "Males," 342 and 357.

56. Court Brown, "Study of Human Sex Chromosome Abnormalities," 394.

57. Court Brown, "Males," 342.

58. Court Brown, "Males," 346.

59. Court Brown, "Males," 348.

60. Court Brown, "Males," 347. Jacobs later echoed Court Brown's reassessment, and especially regretted having used the words *aggressive behavior* in the title of the first brief note in *Nature*. Yet in her defense, she also pointed out that the description of the Carstairs males—whom, as we remember, she did not meet personally—as "dangerous, violent or criminal" corresponded to the definition provided by the institution for its patients; Jacobs, "William Allan Memorial Award Address," 694. In fact, despite regretting the title, Jacobs never quite abandoned the idea of a link between the extra Y chromosome and criminality; see excerpts of an interview with Peter Harper, reproduced in Harper, *First Years*, 90. Jacobs's original study became the object of intense critique, but a later review article by Ernest B. Hook of the Birth Defects Institute at the New York State Department of Health in Albany, New York, fully supported the original observation of an increased incidence of XYY males in medical-penal institutions, providing supporting data from twenty comparable institutions in Europe, North America, and Australia. Hook reported a pooled rate of 2 percent XYY males in these institutions, a figure twenty times the pooled newborn rate; Hook, "Behavioral Implications."

61. Court Brown, "Males," 357.

62. Green, "Media Sensationalisation," 145.

63. Lyons, "Genetic Abnormality."

64. Christitch, "Daniel Hugon"; and Escoffier-Lambiotte, "L'hérédité de la violence." A later article on the matter in the same media outlet commented that the seven-year sentence (from a minimum of five and a maximum of ten years) in the Hugon trial reflected the uncertainty (Fr. *incertitude*) of the jury over how to deal with the genetic evidence; Allain-Régnault, "L'hérédité de la délinquance." On the general resistance of the courts to accept an XYY karyotype as attenuating circumstance, see also Fox, "XYY Offender."

65. "Hidden Perils," *The Times*, 28 July 1967, 3; Eliot Slater, "Genetics of Criminals," *World Medicine*, 21 March 1967, 44–45; Marco Fraccaro, "I figli di Caino," *Il Polso*, 10 February 1969, 1–2; "Of Chromosomes and Crime," *Time*, 5 March 1968, 45–46; "Born Bad?" *Newsweek*, 6 May 1968, 29; "Nature or Nurture?" *New York Times*, 23 April 1968, 46; "The XYY and the Criminal," *New York Times Magazine*, 20 October 1968, 30; Lennard Bickel, "The A-B-C of the X-Y Factor," *The Australian*, 11 October 1968, 9.

66. Statistics showed that violent crimes were on the rise and formed a pervasive aspect of American public life. Intriguingly, Scotland appeared as a distant second after the United States in statistics on murder rates in the late 1960s; see Washington, "Born for Evil?" 328.

67. For an illustration of all these points, including statistical figures on vio-

lent crime, see the various contributions in Mark and Ervin, *Violence and the Brain*. The publication of a four-hundred-page *Bibliography of Aggressive Behavior* in 1977 further illustrates the "explosion of literature" on the subject; Crabtree and Moyer, *Bibliography*. For a historical discussion of the topic, see Durant, "Beast in Man," and Allen, "Modern Biological Determinism" as well as Milam, "Men in Groups"; Milam, *Creatures of Cain*. On the complex interplay of popular and scientific discourse in notions of innate human aggression in the 1960s, see Weidman, "Popularizing the Ancestry of Man." For a more general discussion on the search for biological determinants of human behavior, including a brief discussion of the XYY case, see Gould, *Mismeasure of Man*.

68. Court Brown, "Sex Chromosomes and the Law."

69. See, e.g., Lyons, "Genetic Abnormality."

70. Jacobs, "William Allan Memorial Award Address," 693–94.

71. Engel, "Guest Editorial," 124.

72. Engel, "Guest Editorial," 125.

73. Green, "Media Sensationalisation," 134.

74. Green, "Media Sensationalisation," 147. Unfortunately, the *New York Times* reference given by Green for this quote is incorrect. The original source for the quote could not be traced. It is here reproduced with a caveat.

75. Lejeune, "William Allan Memorial Award Lecture."

76. Green, "Media Sensationalisation," 146.

77. The photo, credited to staff photographer John Oldenkamp, carried no caption.

78. Montagu, "Chromosomes and Crime," 48. This statement, in fact, fit with Montagu's earlier pronouncements on the "natural superiority of women." Although usually read as a provocative argument against biological determinism, Montagu explained the superiority of women by the fact that they were endowed with "two well-appointed, well-furnished X chromosomes"; Montagu, *Natural Superiority of Women*, 130.

79. See Ashley Montagu Papers, Ms. Coll. 109, box 62, folder Chromosomes and Crime, correspondence 1968–1970, APSL.

80. MacLean, Harnden, and Court Brown, "Abnormalities of Sex Chromosome Constitution"; MacLean, "Sex Chromatin Surveys," 203. Unlike the testing of prisoners and other groups of institutionalized individuals, the testing of newborns was still relatively new at the time. The screening programs relied on the approval of the medical heads of the maternity wards. No consent was sought from the babies' parents; see p. 60 in this volume.

81. Court Brown et al., *Abnormalities*. On the registry, see pp. 58–59 in this volume.

82. Ratcliffe, Murray, and Teague, "Edinburgh Study."

83. Interview with Shirley Ratcliffe, Edenbridge, Kent (UK), 1 June 2010.

84. Shirley G. Ratcliffe to Dr. E. G. Sloan, 22 September 1977, file XYY Con-

troversy, Ratcliffe Personal Papers, Edenbridge, Kent. The papers are now held at the Wellcome Library.

85. The studies are listed in order of their starting date. The duration dates for the screening studies of consecutive births were as follows: Denver (January 1964–1974), Edinburgh (April 1967–June 1979), New Haven (October 1967–September 1968), Toronto (October 1967–September 1971), Aarhus (October 1969–January 1974, October 1980–January 1989), Winnipeg (February 1970–September 1973), and Boston (April 1970–November 1974).

86. For details on the study procedure and the evolving consent rules as well as for a comprehensive review of the issues at stake in the Boston study controversy, see Roblin, "Boston XYY Case." For a review of the XYY debate, especially as it developed in the United States, see Hastings Center, "Special Supplement: The XYY Controversy."

87. On Science for the People and a broader analysis of science activists movements and their critique of American science at the time, see K. Moore, *Disrupting Science.*

88. Krimsky, *Genetic Alchemy*, 298–309; Wright, *Molecular Politics*, 37–39.

89. Jensen, "How Much Can We Boost IQ."

90. The study resumed after a review of the consent procedures and was completed in 1973. The results showed no increased incidence of an XYY karyotype in the tested youngsters, an outcome the press did not find newsworthy, thereby missing a significant opportunity to dispel the story of the criminal chromosomes in the public mind; see Washington, "Born for Evil?" Later findings indicated that the XYY karyotype was indeed more frequent in Caucasians than in African Americans; Jensen, "How Much Can We Boost IQ."

91. Money's main interest lay in intersexuality and the proper sex assignment, according to gender (a term created by Money) rather than (biological) sex. For some background on Money's work, see Eder, "Volatility of Sex"; and Eder, "Gender and Cortisone."

92. Beckwith and King, "XYY Syndrome"; Glass, "Science," 23.

93. For an insight into the faculty debate, see the summary report of the meeting: "Faculty Debates Chromosome Study." For Beckwith's account of the story, see Beckwith, *Making Genes, Making Waves*, 125–32.

94. More specifically, funding came from the NIMH Center for Studies of Crime and Delinquency. The cost for screening in the Harvard study was put to six dollars per baby; see Culliton, "Patients' Rights," 716. It appears that some of the chromosome testing was done in conjunction with screening tests for PKU, a metabolic disorder for which a nutritional therapy existed; Kopelman, "Ethical Controversies," 199. Massachusetts became the first state to mandate PKU screening in 1963; Paul and Edelson, "Struggle over Metabolic Screening."

95. Roblin, "Boston XYY Case," 5; Beckwith, "Who Was Wronged?"

96. The issue of informed consent in an era of expanding clinical experimentation was raised several years earlier by Harvard professor of research in anesthesia Henry Beecher in the United States and private medical tutor Maurice Pappworth in the United Kingdom in two publications later regarded as pathbreaking; Beecher, "Ethics"; Pappworth, *Human Guinea Pigs*. However, only the Tuskegee case led to the 1974 Federal Reform Act, which significantly strengthened informed consent rules and instituted institutional review boards in the United States. In the United Kingdom, corresponding legislation was passed only at the end of the decade, and bioethics as a professional endeavor took root in the 1990s; Wilson, *Making of British Bioethics*.

97. Culliton, "Patients' Rights."

98. Beckwith et al., "Harvard XYY Study." A similar position was expressed by Larry Miller, also a member of Science for the People, in a parallel letter in the *Lancet*; L. Miller, "What Becomes of the XYY Male?"

99. Although patients' rights were very much in the center of the debate, those involved in the studies hardly had a voice. An exception is represented by the letter of the mother of a twenty-one-year-old son with an XYY karyotype, published in the *New England Journal of Medicine* in the wake of the termination of the Boston screening program. Against the often-repeated critique that an XYY diagnosis was stigmatizing and harmful, the woman, whose son was tested at age sixteen on her own request, pointed to the "emotional damage of 'not knowing'"—a view quickly countered by Beckwith and others, citing evidence from parents who had participated in the Boston screening study; Franzke, "Telling Parents"; Pyeritz, Beckwith, and Miller, "XYY Disclosure." For a detailed critique of Beckwith's and King's position, see also Hamerton, "Human Population Cytogenetics."

100. McLeod and Johnstone, "Secret Guinea Pigs."

101. Bateman, "Secret Tests."

102. The first volume in the series, featuring the same title as the TV series, appeared in 1970, whereas the television series ran from 1976 to 1977; see Royce, *XYY Man*. The fictional hero of the series, William "Spider" Scott, though intelligent and nonviolent, was unable to leave his criminal past fully behind, presumably because of the extra Y chromosome he carried. Films centered on fictional XYY characters doomed to live criminal lives continued to appear long after the scientific community had laid the image of the aggressive XYY man to rest. See, for instance, the 1992 science-fiction horror film *Alien 3*, directed by American filmmaker David Fincher, evoked by Beckwith; see Beckwith, *Making Genes, Making Waves*, 116.

103. Robinson, Lubs, and Bergsma, *Sex Chromosome Aneuploidy*; Stewart, *Children with Sex Chromosome Aneuploidy*; Ratcliffe and Paul, *Prospective Studies*; Evans, Hamerton, and Robinson, *Children and Young Adults*.

104. Hook, "Geneticophobia."

105. Jacobs, "William Allan Memorial Award Address," 694.

106. Bateman, "Secret Tests," quoting Ratcliffe. Hamerton, in his 1976 presidential address to the American Society of Human Genetics, also strongly pleaded to continue the screening and longitudinal studies to gain more hard data on the issue; Hamerton, "Human Population Cytogenetics."

107. Interview with Shirley Ratcliffe, Edenbridge (Kent), 1 June 2010.

108. Interview with Shirley Ratcliffe, Edenbridge (Kent), 1 June 2010.

109. The question was raised in an editorial of the *British Medical Journal* in 1979; "What Is to Be Done?" (editorial). Nearly ten years earlier, geneticist Bentley Glass, in his speech as outgoing president of the American Association for the Advancement of Science, greeted a near future when "unlimited access to state-regulated abortion will combine with the now perfected techniques of determining chromosome abnormalities in the developing fetus to rid us of the several percentages of all births that today represent uncontrollable defects such as mongolism (Down syndrome) and sex deviants such the XYY type"; Glass, "Science," 28. See also his two rejoinders: Glass, "Less Than Golden Future" and "What Price the Perfect Baby?" A more thoughtful discussion appeared around the same time in a report of the Hastings Center, an independent research institute for bioethics founded in 1969 with headquarters outside New York. Here the XYY case served to illustrate the dilemma of a genetic counselor in presenting "the facts" in a case of "such vague significance as an extra y," followed by a lawyer's response laying out the case of criminal liability when disclosure of even incomplete information is withheld; Veatch, "Unexpected Chromosome"; Annas, "XYY and the Law." According to a Danish survey, all fetuses diagnosed to carry a XYY karyotype were aborted in the country in 1970. The rate fell to 57 percent between 1985 and 1987; Washington, "Born for Evil?" 326. On the dilemma of how to counsel pregnant women in whose fetus an XYY karyotype had been diagnosed and the need for more research on the issue, see also Hamerton, "Human Population Cytogenetics," 119. On prenatal diagnosis for sex chromosome anomalies more generally, see Löwy, *Imperfect Pregnancies*, 123–44.

110. Ratcliffe and Axworthy, "What Is to Be Done?"

111. "The XYY Condition: An Information Sheet to Be Given to Parents and Families" (2005), https://web.archive.org/web/20120222191952/http://www.scotgen.org.uk/documents/XYY%20for%20parents.pdf. Ratcliffe was also involved with the Klinefelter's Syndrome Association UK and responsible for an information booklet on Klinefelter syndrome (47,XXY), issued by the association. A Danish leaflet published in the late 1990s by Dr. Nielsen, a psychiatrist and geneticist who led one of the other seven international newborn screening studies of sex chromosome abnormalities, even more directly addressed the abortion issue and exposed the attitude that XYY fetuses should be aborted as "discriminatory"; "XYY Males: An Orientation," published by the Turner Cen-

ter, Risskov, Denmark, http://web.archive.org/web/20100307091453/http://www
.aaa.dk/turner/engelsk/xyy.htm.

112. Hook, "Behavioral Implications"; Hook, "Extra Sex Chromosomes."

113. Ernest B. Hook, letter to author, 8 December 2008.

114. Interview with John Yates, Cambridge, 15 January 2013. A particularly interesting situation with respect to attitudes toward abortion developed in Glasgow,
where Malcolm Ferguson-Smith, who had done extensive cytogenetic studies in
mental and penal institutions, advocated abortion while his colleague Ian Donald,
the inventor of obstetric ultrasound that made amniocentesis, and with it prenatal
diagnosis, a safer proposition, objected to abortion. According to Ferguson-Smith,
the closer examination of specific cases that he showed his Glasgow colleague
slowly changed Donald's attitude; interview with Malcolm Ferguson-Smith, Cambridge, 28 January 2013. More generally, Donald accepted abortion for what he regarded as grossly abnormal fetuses, but he also purposefully used ultrasound images to deter women from terminating a pregnancy and for pro-life propaganda;
see Nicolson and Fleming, *Imaging and Imagining the Fetus*, 234–49.

115. Montagu, "Chromosomes and Crime," 48.

116. On de Garay and his role in the creation of the Radiobiology and Genetics Department in Mexico City, see Barahona, Pinar, and Ayala, "Introduction
and Institutionalization."

117. de Garay, Levine, and Carter, *Genetic and Anthropological Studies*, xvi.

118. "Sex Tests"; Lyon, "Charles Edmund Ford," 198.

119. Hood-Williams, "Sexing the Athletes," 300.

120. On this point and for an insightful account of the long struggle against
discriminatory gender verification tests, see Ferguson-Smith, "Gender Verification" as well as Simpson et al., "Gender Verification."

121. The introduction of the chromatin test at the Olympics was first proposed by Raymond G. Bunge, an infertility specialist from the University of
Iowa, credited with having founded the world's first sperm bank based on his
invention of preserving sperm in glycerol. Considering the complications of assigning gender in specific cases, he suggested—perhaps jokingly—giving up the
distinction between women and men and instead speaking of the "chromatin-
positive" and "chromatin-negative 100-meter dash"; Bunge, "Sex"; Bunge, "Sex
and the Olympic Games No. 2."

122. Hall, *Girl and the Game*, 158.

123. It remains somewhat unclear which condition exactly led to Klobukowska's disqualification; see Cole, "One Chromosome Too Many?" 129.

124. For a personal testimony by María-José Martínez-Patiño, the athlete
who, in the late 1980s, fought expulsion by the Olympic committee for a failed
chromosome test after having being diagnosed with androgen insensitivity syndrome, see Martínez-Patiño, "Personal Account." Her case helped reshape the

regulations and end chromosome testing at the Olympics. For a discussion of her case, see also Fausto-Sterling, *Sexing the Body*, 1–5.

125. K. L. Moore, "Sexual Identity."

126. Malcolm Ferguson-Smith to James R. Owen, 6 November 1969; J. R. Owen to M. Ferguson-Smith, 11 November 1969; M. Ferguson-Smith to J. R. Owen, 21 November 1969; M. Ferguson-Smith to Col. John Fraser, 21 November 1969, file GB 0248, UGC 188/8/1, Papers of Malcolm Ferguson-Smith, Glasgow University Archives, online on https://wellcomelibrary.org/collections/digital -collections/makers-of-modern-genetics/digitised-archives/malcolm-ferguson -smith/.

127. Ferguson-Smith, "Gender Verification," 361. The fact that androgen insensitivity syndrome is found more frequently in female athletes than in the general female population was explained by the advantage of a higher average stature that accompanies the condition. However, height distinctions (in contrast to sex) are not used as a discriminatory criterion in sports competitions; Ferguson-Smith, "Gender Verification," 364–65. For a critical stance on gender testing, see also de la Chapelle, "Use and Misuse."

128. Interview with Patricia Jacobs, Salisbury, 2 June 2010.

129. Ferguson-Smith, "Gender Verification," 361; and interview with Malcolm Ferguson-Smith, Cambridge, 28 January 2013.

130. Ferguson-Smith, "Gender Verification."

131. The case most frequently discussed in the media is that of the South African middle-distance runner and multiple Olympic gold medalist Caster Semenya, who was banned from international competition in 2009–2010, only to be rehabilitated later while controversy persisted. See McRae, "Semenya Ready." Yet there are other less publicized cases. In May 2019 the International Association of Athletics Federations introduced new rules that set limits to testosterone levels for women athletes, in practice introducing a new gender test. Athletes that exceed these levels are required to take medication to lower their testosterone levels to be able to compete. The ruling is currently under appeal.

132. Nelkin and Tancredi, *Dangerous Diagnostics*.

133. Allen, "Modern Biological Determinism." For a critical take, see Harman, "Unformed Minds."

134. Timmermans and Buchbinder, *Saving Babies?*

Chapter Four

1. de Chadarevian, "Chromosome Photography."

2. On populations as a global political concern in the twentieth century and as a point of intervention for international organizations including the League of

Nations before the war and the United Nations after 1945, see Bashford, "Population"; Bashford, *Global Population.*

3. Caspari, "From Types to Populations"; Lindee and Ventura Santos, "Biological Anthropology of Living Human Populations"; Lipphardt, "Isolates and Crosses"; Bangham and de Chadarevian, "Human Heredity after 1945."

4. Jacobs, "Opportune Life," 36.

5. See pp. 58–59 in this volume.

6. Porter, *Genetics in the Madhouse*; Strasser, *Collecting Experiments*; Stevens, *Life out of Sequence*; Leonelli, *Data-Centric Biology.*

7. For an introduction to recent scholarship on data, see Aronova, von Oertzen, and Sepkoski, *Data Histories*; de Chadarevian and Porter, "Histories of Data and the Database."

8. Group for Research on the General Effects of Radiation, "Proposed Registry of Abnormal Karyotypes," MRC 60/36, file FD 23/142, NA.

9. See W. M. Court Brown, "Proposals for the Future Development of the Clinical Effects of Radiation Research Unit," MRC 67/509, file FD 9/584, NA. Also see interview with Peter Smith, London, 1 September 2010. Freshly graduated in mathematics, Smith worked as a statistician for the Edinburgh unit from the mid- to the late 1960s. For part of this time he was based in the MRC Statistical Research Unit in London, which was directed by Richard Doll, with whom Court Brown collaborated on various projects; see p. 201, note 37, in this volume.

10. The use of Scottish public records for genetic research, including especially the construction of pedigrees, is elucidated in a report by two long-term curators of the Edinburgh registry; Collyer and De Mey, "Public Records."

11. W. M. Court Brown, "Proposal to Report on Cases in the Registry of Abnormal Karyotypes in the Special Report Series" (undated), p. 1, file FD 12/445, NA.

12. Court Brown et al., *Abnormalities*, viii.

13. Collyer and De Mey, "Public Records."

14. W. M. Court Brown, "The Provision of Facilities in Human Cytogenetics to Meet the Needs of Medical Practice," 21 September 1966, file FD 9/1281, NA.

15. Hamerton, "Human Cytogenetic Registries." For a follow-up report, including the description of various registries and their characteristics, see *Human Cytogenetic Registers.* On the Edinburgh meeting, see also J. Miller, "Chromosome Registers."

16. *Medical Research Council Annual Report 2000–2001*, Medical Research Council, London, 2001, pp. 51–52, https://www.mrc.ac.uk/publications/browse/annual-report-and-accounts-200001/.

17. Personal communications, April to December 2011.

18. Personal communications, April to December 2011. On the Scottish mental surveys, see Ramsden, "Surveying the Meritocracy."

19. There was an exception: pedigrees were eventually digitized. In this step

identifying information was excluded; personal communication, December 2011. For an example of how the pedigrees stored in the Edinburgh registry were used to study the effect of chromosomal aberrations on reproductive fitness using computational methods, see Morton et al., "Effect of Structural Aberrations."

20. See W. M. Court Brown, "An Account of the Development of Research in the Medical Research Council's Clinical and Population Cytogenetics Research Unit in Human Cytogenetics and Related Subjects (1956–1969), with Some Observations on Likely Future Progress" (typescript), December 1968, p. 19, MRC Human Genetics Unit, Library and Archive, Edinburgh.

21. W. M. Court Brown to H. Himsworth, 2 June 1964, file FD 23/769, NA.

22. R. C. Norton (MRC) (notes on discussion of Sir Harold and Dr. Court Brown on 16 December 1964), 21 December 1964, file FD 23/769, NA.

23. W. M. Court Brown to H. Himsworth, 2 June 1964, file FD 23/769, NA. The figure corresponds to $2.6 million today.

24. Ledley, "Automatic Pattern Recognition," 2027; Mendelsohn, introduction to *Automation*, 1.

25. November, *Biomedical Computing*, 210–14. From the available records it is difficult to determine with certainty how the collaboration between Ruddle and Ledley started. Apparently Ledley gave a talk at Yale shortly after Ruddle joined the biology faculty in 1961; see interview with Frank Ruddle by Nathaniel Comfort, 4 December 1984, OHHGP. It is likely that Ledley presented his film scanner, and Ruddle, who had done extensive research, including quantitative studies, on chromosomes of the pig, saw the possibility of using the apparatus for the analysis of chromosome images. We can surmise that Ledley developed the necessary software while Ruddle supplied the chromosome images and tested the system on his chromosome preparations. The two researchers shared a National Institutes of Health grant in the early 1960s and, starting in 1964, published a series of papers on their common work; see Ledley, "High-Speed Automatic Analysis"; Ruddle, "Quantitation and Automation"; Ledley and Ruddle, "Automatic Analysis"; Ledley and Ruddle, "Chromosome Analysis." Later Ruddle became instrumental in convening the first Human Gene Mapping Workshop in 1973 and in building up the first human gene mapping database (see chapter 5 in this volume). Barbara Migeon from the Department of Pediatrics at Johns Hopkins also collaborated with Ledley on the automation of chromosome analysis.

26. Ledley and Ruddle, "Automatic Analysis." For the letter of invitation, see J. Faulkner (MRC) to R. S. Ledley, 24 April 1964; box 104, unnamed folder, Ledley Papers, National Library of Medicine, National Institutes of Health, Rockville, MD (NLM).

27. Ledley initially had access to an IBM 7090 computer at the military's classified computer center at the Westinghouse plant in Baltimore; later he used the computer at the Goddard Space Flight Center until he acquired his own

computer in 1967; Ledley, preface to *Research Accomplishments*, xvii–xviii.

28. Ledley agreed to provide the MRC with a copy of the FIDAC instrument for a fixed price of $35,000. The deal later came to haunt him as the NIH's auditors contended that the instrument was built concurrently with work performed under an NIH grant—an accusation that Ledley contested; [R. Ledley], "Discussion of the Position Taken by the Division of Grants Administration Policy of DHEW with Regard to the DCAA [Defense Contract Audit Agency] Audit of the National Biomedical Research Foundation's Books for FY 1965," p. 11 (appended to letter R. Ledley to Senator Mathias, 20 February 1969), unnamed folder, box 104: ADA66-8 Mitotic and Non-Mitotic Cells (box 1 of 2), Ledley Papers, NLM. Together with other accusations of fund mismanagement, this led to a freeze of NIH grants and, ultimately, to a new administrative arrangement for the NBRF in association with Georgetown University. For ample additional material on the audit, including details on the deal with the MRC, see box 104 (uncataloged), Ledley Papers, NLM. Overall, only about a dozen FIDAC scanners were sold to other institutions in the 1960s as the need for a large computer restricted its market; see November, *Biomedical Computing*, 214.

29. W. M. Court Brown to H. Himsworth, 26 November 1964, file FD 23/769, NA.

30. R. C. Norton (MRC) (notes on discussion of Sir Harold and Dr. Court Brown on 16 November 1964), 21 December 1964, file FD 23/769, NA.

31. On the role of the NIH in stimulating the computerization of biology and medicine, including Ledley's role in inspiring that campaign, see November, *Biomedical Computing*. Ledley's pioneering and widely cited volume *Use of Computers in Biology and Medicine* also appeared in 1965. On the situation in Britain, see note 39 in this chapter.

32. "Clinical & Population Cytogenetics Research Unit: Progress by Pattern Recognition Section in Developing Automated Methods of Karyotyping. Draft" [1969], MRC Human Genetics Unit, Library and Archive, Edinburgh.

33. "A Scanner-Computer Combination for Chromosome Studies: An Appreciation and General Specification," May 1966, p. 3 (document for circulation to members of informal meeting at MRC; unsigned but certainly drafted by Rutovitz or Court Brown), file FD 23/157, NA.

34. Jacobs, "Counting," 187. Jacobs made this remark at the same conference at which Ledley first presented his approach to automatic chromosome analysis to a British audience.

35. J. Faulkner, "Note on Visit to Rutovitz with Ministry of Health Representatives," 14 January 1966, and D. Rutovitz, "MRC Clinical Effects of Radiation Research Unit: A First Program of Work for the FIDAC Group," January 1966, p. 8, file FD 23/153, NA.

36. R. C. Norton, "International Symposium on Human Radiation-Induced

Chromosome Aberrations: Future of Court Brown's Unit," 20 October 1966, file FD 23/157, NA.

37. B. L. [Lush] to Dr. Norton, "Need for Further Advice on Automated Karyotyping," 29 September 1966, file FD 23/157, NA.

38. B. L. [Lush], "Visit to Clinical Effects of Radiation Res. Unit, Monday, 13 June," 20 June 1966, p. 2; and R. C. Norton, "International Symposium on Human Radiation-Induced Chromosome Aberrations: Future of Dr. Court Brown's Unit," 20 October 1966, file FD 23/157, NA.

39. "Clinical & Population Cytogenetics Research Unit: Progress by Pattern Recognition Section in Developing Automated Methods of Karyotyping. Draft" (1969), MRC Human Genetics Unit, Library and Archive, Edinburgh. The MRC Computer Centre, which housed a powerful central computer for use by MRC researchers in the London area, was set up in response to the Flowers Committee Report presented to the British Parliament in 1966 to advise on computer needs in research; see "Flowers Report: A Report of a Joint Working Group on Computers for Research," January 1966, http://www.chilton-computing.org .uk/acl/literature/manuals/flowers/foreword.htm; and files FD 23/1714 (Flowers Committee Report and MRC Report in Reply) and FD 23/1726 (Computer Facilities in Central London), NA.

40. Interview with Denis Rutovitz, Edinburgh, 1 April 2008. There is a curious tension between statements by Ledley, who by the mid-1960s regarded the automation of chromosome analysis as solved, and by Rutovitz, who made a more sober assessment of progress; see Piper et al., "Automation," 216. Ruddle, who extensively tested Ledley's system in the mid-1960s, agreed with the Edinburgh group that "many problems remain to be solved"; Ruddle, "Quantitation and Automation," 304. Possibly in response to this critique, Ledley continued to develop his pattern-recognition software. He later described FIDACSYS as a system used "for basic picture handling and object recognition" while the software BUGSYS facilitated "picture analysis and measurement"; Ledley, "Automatic Pattern Recognition," 2017.

41. For more details, see Rutovitz et al., "Computer-Assisted Measurement."

42. Interview with Denis Rutovitz, Edinburgh, 1 April 2008. On military interest in pattern recognition for classifying things in photo surveillance and machine learning around the same period, see Jones, "How We Became Instrumentalists."

43. It is interesting to note that in an analogous field, the automation of cloud chamber readings, the selection of the curves to be measured was done manually, but measuring was automated; see P. M. B. Walker to R. C. Norton, 26 May 1969, file Progress in Pattern Recognition, MRC Human Genetics Unit, Library and Archive, Edinburgh.

44. Piper, "Cytoscan."

45. Piper, "Cytoscan."

46. Interview with Denis Rutovitz, Edinburgh, 1 April 2008. Court Brown died prematurely in 1968, just around the time the Pattern Recognition Group switched to working with the Computer Eye.

47. Mendelsohn, introduction to *Automation*. On the automatic analysis of banded chromosomes and its challenges in the 1970s, see Piper et al., "Automation." An alternative but not necessarily more successful approach to automation was to study chromosomes in flow rather than from a slide, using systems similar to the ones developed for cell sorting. Chromosomes from disrupted cells were kept in suspension, stained with fluorescent dyes, and propelled one by one at high speed through the measuring station of the flow apparatus; see Piper et al., "Automation," 206–7. On the development of flow cytometry for cell sorting, see Keating and Cambrosio, "'Ours Is an Engineering Approach.'"

48. Interview with Denis Rutovitz, Edinburgh, 1 April 2008. On the MRC's increasing interest in technology transfer, see de Chadarevian, "Making of an Entrepreneurial Science."

49. For a comprehensive overview of developments in automated pattern recognition, see Rutovitz, "Reflections on the Past."

50. Jacobs, "Opportune Life," 35–36.

51. Newborn screening for PKU, a hereditary condition that, if left untreated, leads to severe cognitive impairment, is often cited as having led the way for newborn screening. In fact, the first PKU screening programs using the "Guthrie test," based on a bacterial growth test, were put into place only in the early 1960s. Yet screening programs based on the now mostly forgotten "wet-nappy test" for PKU were adopted by single institutions in both the United States and the United Kingdom in 1957; see Paul and Brosco, *PKU Paradox*, 35–53. Results from a first newborn screening program based on sex chromatin testing were published shortly thereafter; K. L. Moore, "Sex Reversal." The screening conducted by Moore, coinventor of the buccal smear technique for Barr body testing, at the University of Western Ontario in Canada, preceded the Edinburgh study.

52. W. M. Court Brown and P. Jacobs, "Clinical Effects of Radiation Research Unit, Population Cytogenetics: The Necessity for the Development of Karyotype Analysis on an Extensive Scale" (memorandum sent to MRC), 21 June 1966, file FD 23/220, NA. See also pp. 59–62 and 98 in this volume.

53. MacLean, Harnden, and Court Brown, "Abnormalities of Sex Chromosome Constitution."

54. W. M. Court Brown, "Contributions of Human Cytogenetics to Clinical Medicine," 16 March 1967, p. 2, file FD 9/1281, NA.

55. On the use of soldiers, prisoners, institutionalized children, and other vulnerable populations as test subjects for research, see Lederer, *Subjected to Science*. On the prisoner as "model system" for biomedical research, see Comfort,

"Prisoner as Model Organism." On asylums as sites for heredity research, see Porter, *Genetics in the Madhouse.*

56. MacLean et al., "A Survey of Sex Chromatin Abnormalities in Mental Hospitals."

57. Jacobs et al., "Chromosome Surveys in Penal Institutions."

58. According to one estimate, by 1972 the chromosomes of seven hundred thousand individuals had been examined worldwide; Soudek, "Chromosomal Variants," 341. The number grew substantially in the following years with the introduction of karyotyping in prenatal diagnosis and the introduction of banding that turned chromosome analysis into a more powerful research and diagnostic tool. Ford estimated that by 1976 three million individuals had been karyotyped in 625 centers around the world; see interview with Charles Ford conducted by Daniel Kevles, Oxford, 25 June 1982, Kevles Papers, FA497, box 1, folder 11 (Ford II, side 1-12), RAC.

59. Jacobs et al., "Chromosome Studies," 340.

60. Jacobs et al., "Chromosome Studies," 341.

61. An example is Malcolm Ferguson-Smith, who did large-scale cytological studies using the buccal smear technique in prison populations in Glasgow in the late 1950s. As a trained physician he considered it important to do all the sampling himself. Interview with Malcolm Ferguson-Smith, Cambridge, 28 January 2013.

62. W. M. Court Brown, "Observations on the Use of Cytogenetic Techniques in the Study of Irradiated Populations. Memorandum Submitted to the Meeting of Investigators in the Field of Studies in Areas of High Natural Radiation," Rio de Janeiro, 12–15 December 1961, file MHO/PA/259.61, WHO Archives. On the WHO's recognition of radiation exposure as a public health issue, see p. 138 in this volume.

63. "Report of the Consultation on the WHO Chromosome Aberration Program and the Preparation of the Manual on Chromosome Aberration Analysis," Geneva, 8–12 November 1971, file RHL/71.6, WHO Archives.

64. Buckton et al., "C- and Q-Band Polymorphisms in the Chromosomes of Three Human Populations."

65. Lipphardt, "Isolates and Crosses."

66. On the similarity between the methods of human population genetics and epidemiology, see the published interview with American geneticist Jack Schull, longtime collaborator with James Neel on the genetic studies of the atomic bomb survivors in Japan; Sever, "Conversation with Jack Schull." I thank Joanna Radin for drawing my attention to this source.

67. T. Morgan, *Heredity and Sex*, 245; T. Morgan, "Has the White Man More Chromosomes."

68. Guyer, "Note," 722.

69. Painter, "Sex Chromosomes of Man."

70. Urofsky, "'Among the Most Humane Moments,'" 15–19.

71. Kodani, "Three Diploid Chromosome Numbers"; Kodani, "Three Chromosome Numbers"; Kodani, "Supernumerary Chromosome of Man."

72. Ford corresponded on the matter with Levan; see C. Ford to A. Levan, 15 January 1957, and A. Levan to C. Ford, 17 January 1957, Correspondence folders, Albert Levan Papers, Special Collections, Lund University Library and Archives. Later it was suggested that Kodani's unusual chromosome counts could have been due to atomic radiation exposure of the people he sampled. On this point and on Kodani's difficult career as an émigré scientist from Japan in America, including his internment in Manzanar during World War II, see Smocovitis, "Genetics behind Barbed Wire."

73. C. E. Ford, "Human Chromosomes. Living Cell Section, Exhibit Number B. 207," William Lawrence Bragg Papers, file MS WLB 85B/22, Royal Institution, Archive Collections. On the science exhibition at the Brussels World Fair, see Lambilliotte, *Le mémorial officiel*; Schroeder-Gudehus, "Popularizing Science."

74. C. E. Ford, "Human Cytogenetics," 105.

75. Makino and Sasaki, "Study of Somatic Chromosomes." The incident is reconstructed in detail in Hyun, "Making Postcolonial Connections."

76. Harnden, "Human Skin Culture Technique." Similar techniques were developed by other researchers around the same time; see, e.g., J. Edwards, "Painless."

77. Harnden, "Early Studies," 165.

78. Harnden, "Chromosomes," 23.

79. The colleague in question was possibly Andrew Arthur Abbie, professor for anatomy and histology at the University of Adelaide. Originally from Britain, he kept in close contact with British colleagues. From the 1950s, his research interest shifted to the anthropology of Aborigines and he conducted many expeditions to Aboriginal communities in South Australia and the Northern Territory. I thank Warwick Anderson for this suggestion.

80. Harnden, "Early Studies," 165.

81. Bangham and de Chadarevian, "Human Heredity after 1945"; Bangham, "Blood Groups and Human Groups." For an attempt to capture these knowledge dynamics, see also Suárez-Díaz, García-Deister, and Vasquez, "Populations of Cognition."

82. On the distinction and tension between typological and population based thinking and the role of individual organisms in respect to type and population, see Mayr, *Systematics*.

83. The notion of indigenous people, which took root in the postwar years under the auspices of the United Nations in an attempt to protect the identity and rights of local populations affected by colonialism, has its own problems of authentication; see Radin, *Life on Ice*, 110. It is, nevertheless, used here as the gen-

erally accepted term today, unless reference is made to different terms used at the time for the same populations.

84. Barnicot and Travers, "Comparison."

85. Hungerford, Giles, and Creech, "Chromosome Studies of Eastern New Guinea Natives." See also Chandra and Hungerford, "Chromosome Studies of Todas"; Hungerford et al., "Chromosome Studies of the Ainu Population."

86. L. L. Cavalli-Sforza, "Research on African Pygmies" (1966, research report), file G3-181-20 (Grant to Istituto di Genetica, Università di Pavia, Italy, in respect of population genetic studies of the Babinga Pygmies), WHO Archives. See also Cavalli-Sforza et al., "Studies on African Pygmies," 255. The Babinga people whom Cavalli-Sforza studied were distinguished from the Baka, the Efe, and other indigenous groups in the forested areas of western Africa.

87. L. L. Cavalli-Sforza to R. L. Kirk (Chief, Human Genetics Unit, WHO, Geneva), 8 October 1966 and 14 October 1966, file G3-181-20 (Grant to Istituto di Genetica, Università di Pavia, Italy, in respect of population genetic studies of the Babinga Pygmies), WHO Archives.

88. World Health Organization, *Research on Human Population Genetics*, 22–27. On the centrality of refrigerating technologies for "salvage anthropology" and the technical, ethical, and diplomatic complexities of collecting blood for this project, see Radin, "Unfolding Epidemiological Stories"; Radin, *Life on Ice.*

89. World Health Organization, *Second Ten Years*, 230–34.

90. On the fallout debate in the United States, see Hacker, *Elements of Controversy*. On the polyvalent significance of the American Atoms for Peace program, see Krige, "Atoms for Peace"; and Creager, *Life Atomic*, 137–42.

91. de Chadarevian, "Human Population Studies."

92. R. Lowry Dobson to Frota-Pessoa (São Paulo), 10 May 1960, file G3-370-2 (Plan for a Medical Research Program Related to Human Genetics, 1955–1966), WHO Archives. The full passage reads: "WHO is greatly interested in human genetics. . . . This interest I am sure you realise, is by no means confined to radiation genetics although this aspect has, because of the historical development of the WHO programme, received special attention so far. On the contrary, the organisation's interest is the broad field of human genetics, of which radiation genetics is a small, albeit significant part."

93. World Health Organization, *Second Ten Years*, 231–32.

94. World Health Organization, *Research on Human Population Genetics*, 5.

95. For more background on the setting up of these two projects, see de Chadarevian, "Human Population Studies"; and James V. Neel, John A. Fraser Roberts, William J. Schull, and Alan C. Stevenson, "Possible Roles of the World Health Organization in Research in Human Genetics. Report of a Meeting of Drs. Neel, Fraser Roberts, Schull, and Stevenson in the Department of Human Genetics, University of Michigan, from April 28th to April 30th, 1959," file

G3/370/2 (Plan for a Medical Research Program Related to Human Genetics), WHO Archives. For the massive report on Stevenson's WHO-sponsored newborn screening survey that collected data on 420,000 pregnancies in twenty-four hospitals in different countries around the world, see Stevenson et al., "Congenital Malformations." In 1960, Stevenson hired Fraccaro, who came to Oxford after research stints with Penrose's group in London and in Uppsala, to set up a cytogenetics laboratory at Oxford. Yet Stevenson himself took up karyotyping only after retirement to study the effects of radiation treatment. Nevertheless, Stevenson's project had some impact on the development of clinical cytogenetics in Mexico. The young Mexican physician Salvador Armendares Sagrera met Stevenson when the latter came to Mexico to present his international project. Later, Armendares spent two years with Stevenson in Oxford as a doctoral student. On his return to Mexico in 1966, he became instrumental in building up clinical cytogenetics in his country; see Barahona, "Medical Genetics." More such studies are needed to understand the impact of the international projects on clinical and research capabilities, including specifically the establishment of cytogenetic facilities, in other regions.

96. On the establishment of the human genetics program at the WHO, see also R. Lowry Dobson (Chief Medical Officer Radiation and Isotopes) to A. Stevenson, 15 November 1959, doc. A14/370/3, WHO Archives. Human genetics was regarded as "one of the high priority fields" in the medical research program; Erwin Kohn (WHO) to Marcel Florkin (Liège), 20 October 1959, file G3-370-2 (Plan for a Medical Research Program Related to Human Genetics, 1955–1966), WHO Archives.

97. On Neel's study of the Xavante and, more generally, the construction of the Xavante as a study population and their role in shaping the relationship with the researchers, see Dent, "Studying Indigenous Brazil."

98. In his autobiography, published in 1994, Neel remarked that, although "primitive" was "the accepted term" for such populations, he had "slowly come to feel not only that it is unduly pejorative" but also that he had "increasing difficulty defining the societal dividing line between primitive and nonprimitive behavior." Here he referred to the recent atrocities committed during the civil war in the former Yugoslavia. He chose to use the term *tribal groups* instead; Neel, *Physician to the Gene Pool*, 120. I follow actors' terms when referring to historical sources but refer to indigenous people otherwise; see also note 83 in this chapter.

99. J. V. Neel to R. Lowry, 31 May 1962, file G3/522/4 (WHO Scientific Group on Research in Populations of Unusual Genetic Interest), WHO Archives.

100. R. L. Kirk (Chief Human Genetics Unit) to Director General, 27 October 1966 (with handwritten response at bottom), and J. Karefa-Smart (WHO Assistant Director-General) to Deputy Director General, "Memorandum," 31 October 1966, file G3/18/20 (Grant to Istituto di Genetica, Università di Pa-

via, Italy, in respect of population genetic studies of the Babinga Pygmies), WHO Archives.

101. Neel et al., "Studies on the Xavante Indians," 52.

102. De Bont, "'Primitives' and Protected Areas." A parallel debate was fueled by the question of whether "primitive people" belonged to "nature" and for this reason should be protected like fauna and flora or whether, on the contrary, they needed to be excluded from the nature parks that were being created. In the postwar era this debate continued between proponents of the idea of "anthropological reserves" and those who countered that overpopulation of indigenous people in the protected areas would pose a threat to the natural environment. The concerns led to the eviction of indigenous people from conservation areas at a time when anthropologists and geneticists were, at least rhetorically, engaged in a race with time to "salvage" what information could still be gained from the "vanishing" populations; Radin, "Latent Life."

103. Neel, "Study of Natural Selection"; Santos, "Indigenous People"; Lindee, "Voices of the Dead."

104. Radin, "Latent Life"; Radin, *Life on Ice*.

105. World Health Organization, *Research on Human Population Genetics*, 7 and 30. The previous WHO technical report on the subject provided very similar guidelines; see World Health Organization, *Research in Population Genetics of Primitive Groups*. Between the publication of the two technical reports the WHO Scientific Group on Research in Population Genetics of Primitive Groups changed its name to Scientific Group on Research on Human Population Genetics. Although the change in name seems significant, references to "primitive" groups continued to be widespread in the body of the text of the report.

106. See, for instance, the volume of the WHO technical report series dealing with the genetic study of "primitive people" where cytogenetics was listed next to a battery of other approaches. Buccal smears for analysis of sex chromatin and blood culture cells for karyotype analysis for the whole population was considered possible in some cases. A smaller number of skin biopsies was also regarded as desirable; World Health Organization, *Research in Population Genetics of Primitive Groups*, 12–13. See also the follow-up report, World Health Organization, *Research on Human Population Genetics*.

107. Interview with Marco Fraccaro, 30 March 2004, Pavia, and file G3/133/4 (Training Course in Human Tissue Culture), WHO Archives. The initial recommendation for the course came from the WHO Research Advisory Group on Human Genetics. Participants selected for the first training course came from Switzerland, France, Poland, Belgium, Germany, Finland, Greece, Italy, Ireland, Sweden, Austria, Denmark, Yugoslavia, Italy, the Soviet Union, England, Portugal, Bulgaria, Egypt, and Iran. Although generally regarded as a great success, the course director noted in his final report that "heterogeneity of cultural background and scientific interest," next to language difficulties, had posed cer-

tain problems; see M. Fraccaro, "Report on the WHO Laboratory Course on Methods of Human Cell Culture and Cytology" (appended to letter M. Fraccaro to R. Lowry Dobson, 3 November 1960), e-file G3/133/4, pp. 356–60, WHO Archives. In subsequent years, the Anatomical Institute in Basel hosted a series of highly successful informal colloquia on chromosomes, at which an officer from WHO regularly participated; see e-file G3/440/6, WHO Archives. It is not clear whether further WHO training courses in chromosome techniques took place. In the 1960s, other organizations, including UNESCO, set up courses in chromosome techniques.

108. World Health Organization, "Genetics and Your Health."

109. "Errors in the Factory." Almasy produced more than one hundred photo stories for the WHO, traveling to many of its regions. On the visual style of communicating science, including especially genetics, by UN organizations such as UNESCO, in the early postwar era, see Bangham, "What Is Race?"

110. A. W. F. Edwards, "Computers and Genealogies," Cambridge 1967 (typescript), courtesy of Anthony Edwards.

111. Montalenti, who occupied the first Italian chair for genetics, established at the University of Naples, was keenly interested in human population genetics and especially in tracing disease frequencies in particular geographic areas. He established the link between the distribution of thalassemia minor and malaria and identified the lack of an enzyme responsible for sugar metabolism as the main cause of favism. On Montalenti and his role in building up genetics in Italy and forging international cooperation, see De Sio, "Genetics and International Cooperation."

112. On the history of the IBP, including especially the Human Adaptability arm of the program, see Worthington, *Evolution of IBP*, 1–16 (with contributions by, among others, Waddington and J. S. Weiner); and Collins and Weiner, *Human Adaptability*, 1–31. On the projects planned as part of the IBP, see J. S. Weiner, *Guide*. On the IBP and its connection with the International Geophysical Year and other big data projects, see Aronova, Baker, and Oreskes, "Big Science and Big Data." On the Human Adaptability program, including especially the blood collection project, see Radin, *Life on Ice*.

113. On the Wenner-Gren Foundation and its support of anthropology in the decades following World War II, see Lindee and Radin, "Patrons of the Human Experience."

114. Baker and Weiner, *Biology of Human Adaptability*, v.

115. de Chadarevian, "Human Population Studies."

116. Neel and Salzano, "Prospectus," 246. On Neel's Amerindian studies, including later ethical controversies surrounding his work in the Amazon, see Lindee, "Voices of the Dead." The volume in which the essay appeared was co-edited by Salzano, Neel's copresenter at the 1964 IBP preparatory meeting. In later years, Neel himself came to wonder whether, by collecting data from the In-

dians, he and his research colleagues were "only the latest exploiters"; Neel, *Physician to the Gene Pool*, 171. For an interview with Salzano and his work with the American Indians, including his collaboration with Neel, see Marcos Pivetta's interview in the magazine *Pesquisa*, from October 2006, at http://revistapesquisa .fapesp.br/en/2006/10/01/a-geneticist-of-polemic-opinions/. On changing ethical frameworks for research on indigenous people in Brazil, see Santos, "Indigenous Peoples." On Neel's research expeditions to Brazil, see also Radin, *Life on Ice*, 86–117; and Dent, "Studying Indigenous Brazil."

117. Collins and Weiner, *Human Adaptability*, 5–6.

118. Collins and Weiner, *Human Adaptability*, 2.

119. Harrison et al., *Human Biology*, cover; Huxley, "Biological Basis." On earlier uses of the term *human biology* by anthropologists focusing on racial hybridity and environmental adaptations in the Pacific, see Anderson, "Hybridity"; Anderson, "Racial Hybridity." On shifting disciplinary understandings and changing practices in physical anthropology in the post–World War II era, see Washburn, "New Physical Anthropology"; Lindee and Ventura Santos, "Biological Anthropology of Living Human Populations"; Lipphardt, "'Geographical Distribution Patterns.'" The "new anthropology" mostly substituted the term *race* for *ethnicity* or *population*, yet the question of how biology related to these notions persisted.

120. Weiner and Lourie, *Human Biology*, xi. The research guide built closely on the two technical reports on the subject issued by the WHO in 1964 and 1968. Commenting on these reports and the IBP guide, Newton E. Morton, who as a young researcher had worked with Neel on the Atomic Bomb Casualty Commission and in principle appreciated the "precious opportunity to study basic problems of human biology in rapidly disappearing primitive people," sharply criticized what he described as the "'collect now, think later' philosophy" that he saw at work in the WHO and IBP programs in population genetics; Morton, "Problems and Methods," 201. On Morton and Neel, their respective notions of primitiveness, and their different approaches in the study of indigenous people, especially in Brazil, see Santos, Lindee, and Souza, "Varieties of the Primitive."

121. Weiner and Lourie, *Human Biology*, 131–39.

122. Collins and Weiner, *Human Adaptability*, 56–61.

123. Collins and Weiner, *Human Adaptability*, 69–71.

124. Collins and Weiner, *Human Adaptability*, 289.

125. Collins and Weiner, *Human Adaptability*, 258.

126. Hamerton et al., "Chromosome Investigations."

127. Bloom et al., "Chromosome Aberrations," 920.

128. Neel, *Physician to the Gene Pool*, 165. I thank Ricardo Ventura Santos for drawing my attention to this study.

129. Jacobs, "Human Chromosome Heteromorphisms."

130. Jacobs, "Opportune Life," 38. On Hawaii as a highly mixed yet racially

equitable society and its special place in studies of race, see Davis, *Who Is Black?*, 109–13.

131. Today such bands are interpreted as regions with moderately or highly repetitive DNA.

132. Jacobs, "Human Chromosome Heteromorphisms," 265.

133. The author of the Australian study referred quite unapologetically to the field of research as "racial cytogenetics"; Angell, "Chromosomes of Australian Aborigines," 103.

134. Jacobs, "Human Chromosome Heteromorphisms," 266.

135. Jacobs, "Human Chromosome Heteromorphisms," 271.

136. See, e.g., Harrison et al., *Human Biology*, 294.

137. Jacobs, "Population Surveillance."

138. Cavalli-Sforza et al., "Call for a Worldwide Survey"; Santos, "Indigenous People."

139. Reardon, *Race to the Finish*; M'Charek, *Human Genome Diversity Project*.

140. Santos, "Indigenous Peoples"; Kowal, "Orphan DNA"; Kowal and Radin, "Indigenous Biospecimen Collections"; Radin and Kowal, "Indigenous Blood"; World Health Organization, "Indigenous People and Participatory Health Research."

141. Interview with Nick Hastie, Edinburgh, 20 July 2010. Generation Scotland, launched in 2004, is a further genetic database project based in Edinburgh. It includes data of a family-based cohort with the aim of identifying genetic risk factors for common complex diseases. More recently, genomic data has been added to the database. On Generation Scotland, see Reardon, *Postgenomic Condition*, 94–119.

142. Epstein, *Inclusion*.

Chapter Five

1. Hsu, *Human and Mammalian Cytogenetics*, 161.

2. Crick, "Double Helix," 767.

3. Watson, "Foreword," xv.

4. Hsu, *Human and Mammalian Cytogenetics*, 162.

5. Hsu, *Human and Mammalian Cytogenetics*, 162.

6. On the concept of molecularization, see de Chadarevian and Kamminga, introduction to *Molecularizing Biology and Medicine*.

7. Galison, *Image and Logic*.

8. The use of the term *tradition* here is intended to highlight the process of acculturation that is necessary to grow into and share a tradition as well as to the difficulties inherent in bridging research traditions. Fleck's notion of thought

collective, Kuhn's paradigms, Hacking's styles of thought, and Keating and Cambrosio's notion of biomedical platforms all include similar elements; see Fleck, *Genesis*; Kuhn, *Structure*; Hacking, "Styles of Scientific Reasoning"; Keating and Cambrosio, *Biomedical Platforms*.

9. Harper, *First Years*, 163.

10. Ferguson-Smith, "From Chromosome Number to Chromosome Map"; Harper, *First Years*, 163–64.

11. FISH built on earlier in situ hybridization experiments with radioactively labeled DNA or RNA and autoradiographic methods for detection. Although these earlier techniques were cumbersome and limited in their scope, they were greeted enthusiastically by Hsu, who predicted a "marriage between molecular biology and cytology"; Hsu, *Human and Mammalian Cytogenetics*, 108–15.

12. Sumner and Chandley, *Chromosomes Today*, xxiii. The first meeting of what developed into the International Chromosome Conferences was organized by Cyril Darlington in 1964 to bring together scientists working on different aspects of chromosome research. The conferences continue to the present day.

13. Quoted after Olby, *Francis Crick*, 78.

14. Olby, *Francis Crick*, 78.

15. See, e.g., Allen, *Life Science in the Twentieth Century*. For a critical view on the sharp dichotomy between observational, including comparative, and experimental approaches in mid-twentieth-century biology, see de Chadarevian, "Mapping Development"; Strasser and de Chadarevian, "The Comparative and the Exemplary;" Strasser, *Collecting Experiments*.

16. Crick, *Of Molecules and Men*, 13–14.

17. de Chadarevian, "Mapping Development."

18. Speech by S. Brenner (draft) (Royal Society discussion on the proposed laboratory of the European Molecular Biology Organization, 21 October 1969), Kendrew Papers, file F.189, Bodleian Library, Oxford.

19. The available sources do not indicate how the collaboration between Barnicot and Huxley started but most likely Barnicot was the driving force. On his comparative studies of human chromosomes, see p. 137 in this volume.

20. Ferguson-Smith, "Putting Medical Genetics into Practice."

21. Barnicot and Huxley, "Electron Microscopy." Barnicot and Huxley published a couple more articles together describing a method for sectioning mitotic chromosomes for electron microscopic study. Other researchers followed their lead. See, e.g., E. H. R. Ford, Thurley, and Woollam, "Electron-Microscopic Observations"; and, more recently, Saitoh and Laemmli, "Metaphase Chromosome Structure." The lead author of the first paper, the Cambridge anatomist E. H. R. Ford, later wrote a major monograph on human chromosomes; E. H. R. Ford, *Human Chromosomes*.

22. Olby, *Francis Crick*, 351–53.

23. Crick, "General Model," 27.

24. Crick, "Double Helix," 768.

25. Finch, *Nobel Fellow*, 127.

26. Kornberg, "Chromatin Structure."

27. For a fuller account of the history of chromatin research, see Olins and Olins, "Chromatin History." Kornberg went on to study the transcription process by which DNA is transformed into RNA in molecular detail by "freezing" the responsible enzyme, RNA polymerase, in action. For this work he received the Nobel Prize in Chemistry in 2006.

28. Interview with John Edwards by Peter Harper, 23 August 2004, GenMedHist.

29. Dutrillaux et al., "Sequence of DNA Replication."

30. It is interesting to note in this context that Watson's widely used textbook of molecular biology, first issued in 1965, included sections on the chromosomal view of heredity, on the role of chromosomes in mitosis and meiosis, and on the arrangement of genes in viral and bacterial chromosomes. Yet nowhere did the text mention the number of human chromosomes or included a representation of the human karyotype. This topic clearly did not belong into the new canon of molecular biology; Watson, *Molecular Biology of the Gene*.

31. Crick, "Postscript," 406.

32. Lacadena, "Cytogenetics," 9.

33. Kohler, *Lords of the Fly*.

34. Kohler, *Lords of the Fly*. The extent to which *Drosophila* genetics set the standard for the genetics of other organisms, including humans, can be gleaned from the fact that, well into the 1950s, sex determination in humans was expected to work as in the fly (see p. 48 in this volume). On the significant contributions of British researchers to the statistical analysis of genetic linkage in the 1930s, see A. Edwards, "Linkage Methods."

35. Rheinberger and Gaudillière, *Classical Genetic Research*; Gaudillière and Rheinberger, *From Molecular Genetics to Genomics*.

36. Holmes, "Seymour Benzer and the Definition of the Gene"; Holmes, "Seymour Benzer and the Convergence of Molecular Biology"; Holmes and Summers, *Reconceiving the Gene*. On Benzer, see also J. Weiner, *Time, Love, Memory*.

37. de Chadarevian, "Mapping Development."

38. Haldane, "Provisional Map"; Bell and Haldane, "Blindness and Haemophilia."

39. Hogben, *Genetic Principles*, 214.

40. Mazumdar, *Eugenics, Human Genetics and Human Failings*, 166–69; Kevles, *In the Name of Eugenics*, 193–98; Schneider, "Blood Group Research"; Schneider, "History of Research on Blood Group Genetics"; Bangham, "Blood Groups and the Rise of Human Genetics." Geographical mapping of blood group frequencies in the context of human diversity studies preceded the use of

blood groups as markers for chromosome linkage mapping; Gannett and Griese-mer, "ABO Blood Groups."

41. The first autosomal linkage between a blood group (the Lutheran blood group) and another human gene (for a human secretory factor) was established in 1951. The genes were later shown to be located on the long arm of chromosome 19; see V. McKusick, "Twenty-Five Years of Human Genome Meetings (HGMs): The Past and the Future (Draft 1)," 1 April 1998 (typescript), Victor McKusick Papers, box 509632, file HX, Hopkins Medical Archives.

42. Polani, "Human and Clinical Cytogenetics," 118.

43. Haldane, "Formal Genetics of Man," 149.

44. On McKusick's work with the Amish and the construction of pedigrees, see Lindee, *Moments of Truth*, 58–89. On the collaboration between Renwick and Schulze on the first computer linkage program and the eventual unraveling of the international collaboration, see Victor McKusick Papers, file Renwick, Dr. James H., box 509328, Hopkins Medical Archives; and J. H. Renwick, "Computing the Location of Genes in Man" and other documents in James Renwick Papers, folder Renwick Computers, online at http://wellcomelibrary.org/player/b20104868. McKusick's work with computers was featured in an article in IBM's house journal, *Computing Report for the Scientist and the Engineer*; see "Tracing Genetic Diseases." On the harnessing of computers for human gene mapping and the cataloging of genetic diseases at the Moore Clinic in Baltimore, see also McGovern, "'The London/Baltimore Link.'"

45. McKusick, "60-Year Tale," 10.

46. Donahue et al., "Probable Assignment."

47. Caspersson, Lomakka, and Zech, "24 Fluorescence Patterns."

48. For instance, French researchers found that heating the chromosome preparations in hot phosphate buffer before Giemsa staining produced a reversed banding pattern to that of G-bands: dark G-bands were stained lightly, and vice versa. R-banding was especially useful for the study of the distal ends of chromosomes that tend to be more darkly stained with this technique. Often R- and G-bands were used in conjunction.

49. Hsu, *Human and Mammalian Cytogenetics*, 123.

50. Hsu, *Human and Mammalian Cytogenetics*, 101 and 127.

51. Hamerton, Jacobs, and Klinger, "Paris Conference"; Hamerton and Klinger, *Paris Conference (1971)*. On the human chromosome standardization conferences, see pp. 63–74 in this volume. In the mid-1970s, cytogeneticists moved from metaphase to prophase chromosome preparations, as the more elongated chromosomes in the earlier stage of cell division showed a much larger number of bands than the best metaphase preparations. This move was connected with a change in the representation of chromosomes from X-shaped (showing the early separation of two daughter chromosomes) to rod-shaped

(prophase chromosomes). The step facilitated the physical mapping of genes on the chromosomes; see Hogan, "'Morbid Anatomy.'" In contrast to linkage maps that represent the relative distance between genes, physical maps indicate the actual (physical) location of genes on chromosomes.

52. A prominent example is the use of cell fusion for the production of monoclonal antibodies, a technique pioneered by César Milstein and Georges Köhler in the mid-1970s. In their experiments the two researchers fused mouse cancer cells with mouse antibody-producing cells. On the commercialization and clinical use of the technique, see Cambrosio and Keating, *Exquisite Specificity*; de Chadarevian, "Making of an Entrepreneurial Science"; Marks, *Lock and Key of Medicine*. On somatic cell hybrids and gene mapping, see Harris, *Cells of the Body*; Ferguson-Smith, "From Chromosome Number to Chromosome Map"; and Polani, "Human and Clinical Cytogenetics." On the new notions of biological hybridity that accompanied the novel experimental techniques, see Landecker, *Culturing Life*, 180–218.

53. O. Miller et al., "Human Thymidine Kinase Gene Locus."

54. Ruddle et al., "Linkage Relationships."

55. Harris, *Cells of the Body*, 136. By the time Ruddle convened the first International Workshop on Human Gene Mapping in New Haven in 1973, thirty-three genes had been mapped on eighteen chromosomes using the same technique, with the number climbing rapidly. On the coinage of the term *somatic cell genetics*, see interview with Frank Ruddle by Nathaniel Comfort, 4 December 1984, p. 19, OHHGP. I was unable to confirm the information independently.

56. "New Haven Conference," 66.

57. "New Haven Conference," 9–10.

58. V. McKusick, "Twenty-Five Years of Human Genome Meetings (HGMs): The Past and the Future (Draft 1)," 1 April 1998 (typescript), pp. 1 and 7, Victor McKusick Papers, box 509632, file HX, Hopkins Medical Archives.

59. Ferguson-Smith, "From Chromosome Number to Chromosome Map," 11, table 2.

60. "New Haven Conference," 21. For some background on the workshops, see also Jones and Tansey, *Human Gene Mapping Workshops*.

61. Pardue and Gall, "Molecular Hybridization." It is not easy to tease apart who brought which skill to the collaboration between Gall and Pardue; see Hsu, *Human and Mammalian Cytogenetics*, 109–15. Both studied biology in the 1950s, specializing in cell biology, and developed an interest in chromosomes. Eventually they both embraced molecular techniques. Pardue received her PhD from Yale in 1970, concurrent with her collaboration with Gall. The latter participated at the first Human Gene Mapping Workshop at Yale in 1973 but was not among the speakers. Both participated but did not present papers at the Cold Spring Harbor Symposium on Chromosome Structure and Function convened by Watson in 1973.

62. Hirschhorn and Boyer, "Report of the Committee on In Situ Hybridization," 55.

63. Berg, "Introduction," 2.

64. McKusick, "Human Genome," 15.

65. McKusick, "Human Genome," 6.

66. Southern, "Application of DNA Analysis," 52 and 53.

67. Botstein et al., "Construction of a Genetic Linkage Map"; Skolnick and Francke, "Report of the Committee on Human Gene Mapping"; Ruddle and Kidd, "First Human Gene Mapping Interim Meeting."

68. Gusella et al., "Polymorphic DNA Marker." It took ten more years to identify the mutant gene. For a personal historical account of the hunt for the gene, see Wexler, *Mapping Fate.*

69. Ruddle and Kidd, "First Human Gene Mapping Interim Meeting."

70. Ruddle and Kidd, "Human Gene Mapping Workshops," 1.

71. Advisory Committee to the Director, NIH, Minutes of 54th Meeting, 16–17 October 1986, box 0103-003. National Human Genome Research Institute, Archival and Digitized Materials, Bethesda, MD (NHGRI Archive).

72. Howard Hughes Medical Institute Human Gene Mapping Library (9/1988), box 0148A-006, NHGRI Archive.

73. Ruddle and Kidd, "Human Gene Mapping Workshops." Mapping here did not relate to gene mapping but rather to the construction of a physical map of the genome composed of overlapping DNA fragments as a first step toward sequencing.

74. Solomon and Bodmer, introduction to "Human Gene Mapping 11"; Probert and Rawlings, "Overview"; Rawlings et al., "Report."

75. On this point, see also Theodore Puck, "Memorandum on Dr Betinsky's [sic] Human Genome Conference" (Santa Fe meeting), attached to letter T. Puck to M. Betinsky [sic], 17 March 1986, box 0102-001, NHGRI Archive.

76. On Ruddle's participation at the meeting in Santa Fe, see material in box 0102-001 and box 0103-001, NHGRI Archive. On Ruddle's reflections on the workshop and his vision for the sequencing project, see especially Ruddle to Bitensky, 17 March 1986, box 0102-001. On his participation in the committee that presented the Human Genome Project to Congress, see interview of Frank Ruddle by Dmitriy Myelnikov, New Haven, CT, 8 December 2011. I thank Dmitriy Myelnikov and Nancy Ruddle for making the interview available to me.

77. McKusick and Ruddle, "Editorial," 1.

78. On Ruddle as possible candidate for the directorship, see, e.g., C. Thomas Caskey to James Wyngaarden (Director, NIH), 2 March 1988, box 0102-008, NHGRI Archive.

79. On the perceived problems with the Yale database and its discontinuation, see material in box 0148A-006, NHGRI Archive.

80. Interview with Kenneth Kidd, New Haven, CT, 12 May 2016. In 1992 GenBank was moved from Los Alamos to the National Center for Biotechnol-

ogy Information, housed on the NIH campus in Bethesda, MD as part of the National Library of Medicine. On the history of GenBank and the moral economy that governed it, see Strasser, "Experimenter's Museum."

81. Interview with Frank Ruddle by Nathaniel Comfort, 4 December 1984, p. 34, OHHGP. On the technical contributions of somatic cell genetics to the Human Genome Project and molecular biology more generally, see also Harris, *Cells of the Body*, 153–209; and Hogan, *Life Histories*, 56–86.

82. V. McKusick, "Twenty-Five Years of Human Genome Meetings (HGMs): The Past and the Future (Draft 1)," 1 April 1998 (typescript), pp. 6 and 10, Victor McKusick Papers, box 509632, file HX, Hopkins Medical Archives.

83. On the problems of defining what a gene is up to proposing the abolition of the term, see Keller, *Century of the Gene*; Beurton, Falk, and Rheinberger, *Concept of the Gene*; and Rheinberger and Müller-Wille, *Gene*.

84. Hogan, *Life Histories*, 186–96.

85. Historical reasons militated against the inclusion of human genetics into molecular biology. This was especially the case in Germany, as illustrated by the history of the Max Planck Institute for Molecular Genetics in Berlin. The institute was founded in the mid-1960s with the clear intention of marking a break with its predecessor, the Max Planck Institute for Comparative Genetics and Hereditary Pathology, which traced its origins to the deeply tainted Kaiser Wilhelm Institute for Anthropology, Human Genetics, and Eugenics. The newly founded institute was to be solely based on "pure" research on cells, viruses, and bacteria, excluding all applications in human genetics. The intention to exclude human genetics in the new institute was contested. Opponents pointed out that the field had been gaining new currency, not least because of its relevance in respect to the development of nuclear energy; see Sachse, "Ein 'als Neugründung zu deutender Beschluss.'" Nevertheless, human genetics remained off-limits at the Berlin institute until the mid-1990s, when, after a highly polarized debate, a reorientation of the institute toward the analysis of human and other genomes, disease causation, and medical treatments was agreed; Trautner, "'Ich hätte mir gar nichts anderes vorstellen können,'" 70–71; Sperling, "50 Jahre"; see also the website "Max Planck Institute for Molecular Genetics—History," at http://www.molgen.mpg.de/3498/Geschichte.

86. Interview with John Yates, Cambridge, 15 January 2013. Yates, who entered the field in the early 1980s, trained first at Edinburgh and then at Glasgow under Malcolm Ferguson-Smith. He later moved with Ferguson-Smith to Cambridge to build up medical genetics there.

87. McKusick, "60-Year Tale."

88. Interview with John Yates, Cambridge, 15 January 2013; Anthony W. F. Edwards, personal communication, 8 October 2012. The situation was different in the London and Glasgow area (see chapter 2 in this volume).

89. Morange, *History*.

90. For exceptions, see Comfort, *Science of Human Perfection*, chap. 7; and Hogan, *Life Histories*, 121.

91. de Chadarevian, *Designs for Life*, 345–53.

92. Fortun, "Projecting Speed Genomics"; Hilgartner, "Constituting Large-Scale Biology"; Hilgartner, *Reordering Life*.

93. See pp. 62–63 in this volume.

94. Interview with John Yates, Cambridge, 15 January 2013.

95. Ferguson-Smith, "From Chromosome Number to Chromosome Map," 16. In his work on the evolution of the chromosomes of vertebrates Ferguson-Smith continued to combine molecular techniques such as FISH with cytogenetic observations; see Ferguson-Smith, "Putting Medical Genetics into Practice."

96. Interview with Ann Chandley, Edinburgh, 21 July 2010.

97. Parallel to heading the new Molecular Genetics Section in the Clinical and Population Cytogenetics Unit, Southern also directed the MRC Mammalian Genome Unit in Edinburgh. Founded in 1972, the unit was dedicated to the study of the structure and function of chromosomes on the molecular level. The arrangement, orchestrated by the MRC, was seen as serving both units by providing molecular expertise to the cytogeneticists and clinical applications to the molecular biologists. On Southern's move to Oxford in 1985, the Mammalian Genome Unit was absorbed into the Clinical and Population Cytogenetics Unit. The Mammalian Genome Unit is not to be confused with the Mammalian Genetics Unit at Harwell that was created in the mid-1990. On Southern and molecular biology in Edinburgh, see Martynoga, *Molecular Tinkering*.

98. This early molecular mapping project preceded the Human Genome Mapping Project proposed by Cambridge-based molecular biologist Sydney Brenner in 1986 and officially launched by the MRC in 1989 with the establishment of the Human Genome Mapping Resource Centre at Hinxton, at the outskirts of Cambridge. A few years later, the Sanger Centre (now Sanger Institute) that made a decisive contribution to the international human genome sequencing project, was built on the same site. Brenner's mapping project was eventually also subsumed under the international Human Genome Project. It is noteworthy that clinically inclined cytogeneticists, including, for instance, Malcolm Ferguson-Smith, were actively involved in the discussions on Brenner's mapping project; see García-Sancho, "Proactive Historian."

99. "Medical Research Council Clinical and Population Cytogenetics Unit, Western General Hospital, Edinburgh. Progress Report February 1979–June 1982." MRC Human Genetics Unit, Library and Archive, Edinburgh.

100. Interview with Ann Chandley, Edinburgh, 21 June 2010.

101. Interview with Ann Chandley, Edinburgh, 21 June 2010.

102. From notes taken at interview with Nick Hastie, Edinburgh, 20 July 2010.

103. Historians have drawn attention to two further examples in which cytogenetic observation in tandem with clinical studies paved the way for a molecu-

lar understanding of the underlying disease mechanism and the description of novel genetic principles. The first concerns the study of chromosome translocation in chronic myeloid leukemia and its impact on the oncogene theory of cancer; the second regards the cytogenetic study of fragile X syndrome and the insights this work generated into the effects and the heritability of genomic instabilities; see Keating and Cambrosio, "New Genetics and Cancer"; and Hogan, "Disrupting Genetic Dogma." Both studies highlight the (neglected) contribution of cytogenetics to the making of late twentieth-century biomedicine. The investigation of telomeres, the ends of chromosomes, and their role in the ordered replication of the genetic material and in aging, is a further example, but here research was based on chromosomes from a variety of organisms.

104. Quoted from notes taken from a conversation with Wendy Bickmore, Edinburgh, 20 July 2010. See also interview with Wendy Bickmore, Edinburgh, 11 July 2017. In 2017, Bickmore became director of the Edinburgh unit, now renamed the MRC Human Genetics Unit.

105. See, e.g., Cremer and Cremer, "Chromosome Territories." The review article includes an overview of the "evolving toolkit for studies of nuclear structures" that link microscopy and molecular biology. On the new interest in the packing structure of DNA as a three-dimensional body that forms the hub for epigenetic signaling networks, see also Landecker, "Social as Signal."

106. Interview with Sibel Kantarci and Rao Nagesh, UCLA Clinical and Cytogenetics Laboratory, 24 April 2014. I thank Kantarci and Wayne Grody (UCLA) for providing me with a copy of the cartoon. Kantarci first trained as a molecular geneticist (2004 Harvard) but then felt she "missed the big picture" and added training in cytogenetics (2007). She holds a double certification as a clinical molecular geneticist and a clinical cytogeneticist. On the continuing dominance of the microscope over molecular technologies in cancer pathology more generally, see Hogarth, Hopkins, and Rodriguez, "Molecular Monopoly?" esp. 246. On resistances to the "molecular turn" and the persistence of cytogenetic techniques in prenatal diagnosis, see Turrini, "Controversial Molecular Turn." On the view that molecular techniques complemented rather than replaced chromosomal techniques in chromosome mapping and medical analysis, see also Hogan, Life Histories, chap. 4.

107. Interview with Rao Nagesh, UCLA Clinical and Cytogenetics Laboratory, 24 April 2014. On the history of Gleevec and the molecularization of cancer therapy more generally, see also Keating and Cambrosio, Cancer on Trial, 315–45.

108. Peter Keating, "Why Is Gleevec a Paradigm for Targeted Therapy?" (unpublished manuscript). I thank Peter Keating for making the manuscript available to me.

109. Rheinberger, Toward a History.

110. Crick, "Postscript," 403.

111. Lima-de-Faria, One Hundred Years; Lima-de-Faria, Praise.

112. Malcolm Ferguson-Smith, editorial reviews at http://www.amazon.com/Hundred-Chromosome-Research-Remains-Learned/dp/1402014392.

113. Kay, *Who Wrote the Book of Life?*

114. Anker and Nelkin, *Molecular Gaze*, 17–45.

115. Gilbert, "Vision of the Grail," 96; de Chadarevian, "Chromosome Photography," 139–41.

116. The reading of visual patterns in maize kernels on which Barbara McClintock based her work on mobile genetic elements has been presented as a barrier to communication with molecular biologists; see Keirns, "Seeing Patterns." However, there are multiple examples—from X-ray diffraction patterns to readings of microarrays—where pattern analysis has played a role in molecular research. There are surprising convergences of methods as well. For instance, the computer software employed to read chromosome banding patterns and gel electrophoresis bands—a widely used method for the separation and analysis of macromolecules—is based on the same principle.

117. Taylor, Woods, and Hughes, "Organization and Duplication of Chromosomes"; Taylor, "Sister Chromatid Exchanges"; Meselson and Stahl, "Replication of DNA." Semiconservative replication implies that one strand of the parent DNA material is conserved intact in each of the two daughter cells. On "the most beautiful experiment in molecular biology," including reference to Taylor's work, see Holmes, *Meselson, Stahl, and the Replication of DNA*.

118. Creager, *Life Atomic*, 3–4.

119. Confirming the salience of the new imaging techniques, several recent Nobel Prizes went to the developers of the new techniques. Martin Lee Chalfie, Osamu Shimomura and Roger Y. Tsien shared the 2008 Nobel Prize in Chemistry for the discovery and development of the green fluorescent protein and the 2014 chemistry prize was shared by Eric Betzig, Stefan W. Hell, and William E. Moerner for the development of super-resolved fluorescence microscopy. One of the prize recipients explicitly remarked that the goal of developing the new microscope was "to link the fields of molecular and cellular biology"; Claudia Dreifus, "Eric Betzig's Life over the Microscope," *New York Times*, 28 August 2015, http://www.nytimes.com/2015/09/01/science/eric-betzig-life-over-the-microscope.html.

120. Amos and White, "How the Confocal Laser Scanning Microscope Entered Biological Research."

121. See, e.g., Yu et al., "Probing Gene Expression."

122. Interview with Wendy Bickmore, Edinburgh, 11 July 2017.

123. On radioactive tracers and scintillation counters as a key technology for molecular biology, see Rheinberger, "Putting Isotopes to Work"; and Creager, "Phosphorus-32"; on centrifuges, Elzen, "Two Ultracentrifuges"; on X-ray diffraction, de Chadarevian, *Designs for Life*; on electron microscopes, Rasmussen, *Picture Control*.

124. Landecker, "Life of Movement," 395.

Epilogue

1. Harnden, "Early Studies."

2. Poole, *Earthrise*.

3. Lindee, "Genetic Disease," 1.

4. Duster, *Backdoor to Eugenics*.

5. Benjamin, "Lab of Their Own."

6. Kowal and Radin, "Indigenous Biospecimen Collections"; Radin and Kowal, "Indigenous Blood."

7. Hogan, *Life Histories*.

8. Daston and Lunbeck, "Introduction." In John Pickstone's scheme of different "ways of knowing" observations under the microscope belong to the "natural historical" and "analytical" rather than the "experimental" way of knowing. Since the nineteenth century the latter has claimed more authority, yet earlier ways of knowing persist; Pickstone, *Ways of Knowing*, 146. On the opposition, from the nineteenth century onward, of "active" experiment and observation, presented as the passive registration of data, see also Daston and Lunbeck, "Introduction."

9. Daston and Galison, *Objectivity*, 253–307. Nor is the distrust in the visual confined to the sciences. On the denigration of vision and its connection to the project of modernity in French epistemology, see Jay, *Downcast Eyes*. On the cultural appeal of quantitative methods as a measure for objectivity, see Porter, *Trust in Numbers*.

10. Crick, *What Mad Pursuit*, 67.

11. See p. 70 and p. 218, note 117, in this volume.

12. Interview with Wendy Bickmore, Edinburgh, 11 July 2017. On the difference between seeing with the eye and with the computer, see also Pitt, "Epistemology of the Very Small."

13. Lade, "Une belle image."

14. Stevens, *Life out of Sequence*, 171–201.

15. A somewhat similar account has been presented for the relation of particle and solid state physics. Despite their mutual dependence, particle physics marshaled more prestige and public attention during the Cold War than solid state physics that for its links to industry was portrayed as "dirty physics" (or "Schmutzphysik"); see Joseph D. Martin, *Solid State Insurrection*.

16. Keller, *Century of the Gene*.

17. Richardson and Stevens, *Postgenomics*; Rheinberger and Müller-Wille, *Gene*.

18. Keller, "Postgenomic Genome."

Bibliography

Adinolfi, Matteo, Philip Benson, Francesco Giannelli, and Mary Seller, eds. *Paediatric Research: A Genetic Approach. Festschrift for Paul Polani.* London: Spastics International Medical Publications, 1982.

Allain-Régnault, Martine. "L'hérédité de la délinquance: La criminologie n'est pas un moyen d'excuser le phénomène criminel." *Le Monde,* 6 June 1972, 12.

Allen, Garland E. *Life Science in the Twentieth Century.* Cambridge: Cambridge University Press, 1978.

———. "Modern Biological Determinism: The Violence Initiative, the Human Genome Project, and the New Eugenics." In *The Practices of Human Genetics,* edited by Michael Fortun and Everett Mendelsohn, 1–23. Dordrecht, The Netherlands: Kluwer Academic Publishers, 1999.

Allen, Gordon, et al. "Mongolism (Correspondence)." *Lancet* 277, no. 7180 (1961): 775.

Amos, W. B., and J. G. White. "How the Confocal Laser Scanning Microscope Entered Biological Research." *Biology of the Cell* 95 (2003): 335–42.

Anderson, Warwick. "Hybridity, Race, and Science: The Voyage of the *Zaca,* 1934–1935." *Isis* 103 (2012): 229–53.

———. "Racial Hybridity, Physical Anthropology, and Human Biology in the Colonial Laboratories of the United States." *Current Anthropology* 53 (2012): S95–S107.

Angell, Roslyn. "The Chromosomes of Australian Aborigines." In *The Human Biology of Aborigines in Cape York,* edited by R. L. Kirk, 103–9. Australian Aboriginal Studies 44. Human Biology Series 5. Canberra: Australian Institute of Aboriginal Studies, 1973.

Anker, Suzanne, and Dorothy Nelkin. *The Molecular Gaze: Art in the Genetic Age.* Cold Spring Harbor, NY: Cold Spring Harbor Laboratory Press, 2003.

Annas, George Joseph. "XYY and the Law" (letter). *Hastings Center Report* 2, no. 2 (April 1972): 14.

Appadurai, Arjun, ed. *The Social Life of Things: Commodities in Cultural Perspective*. Cambridge: Cambridge University Press, 1986.

Arnold, Lorna. *Windscale 1957: Anatomy of a Nuclear Accident*. New York: St. Martin's Press, 1992.

Aronova, Elena, Karen S. Baker, and Naomi Oreskes. "Big Science and Big Data in Biology: From the International Geophysical Year through the International Biological Program to the Long-Term Ecological Research (LTER) Network, 1957–Present." *Historical Studies in the Natural Sciences* 40 (2010): 183–224.

Aronova, Elena, Christine von Oertzen, and David Sepkoski, eds. *Data Histories*. Vol. 32 of *Osiris*. Chicago: University of Chicago Press, 2017.

Auerbach, Charlotte. *Genetics in the Atomic Age*. Edinburgh: Oliver and Boyd, 1956.

Awa, A. A., A. D. Bloom, M. C. Yoshida, S. Neriishi, and P. G. Archer. "Cytogenetic Study of the Offspring of Atom Bomb Survivors." *Nature* 218 (1968): 367–68.

Awa, A. A., T. Honda, S. Neriishi, T. Sufuni, H. Shimba, K. Ohtaki, et al. "Cytogenetic Study of the Offspring of Atomic Bomb Survivors, Hiroshima and Nagasaki." In *The Children of Atomic Bomb Survivors: A Genetic Study*, edited by James V. Neel and William J. Schull, 344–61. Washington, DC: National Academy Press, 1991.

Baker, Paul T., and Joseph Sidney Weiner, eds. *The Biology of Human Adaptability*. Oxford: Clarendon Press, 1966.

Bangham, Jenny. "Blood Groups and Human Groups: Collecting and Calibrating Genetic Data after World War II." *Studies in History and Philosophy of Biological and Biomedical Sciences* 47A (2014): 74–86.

———. "Blood Groups and the Rise of Human Genetics in Mid-Twentieth Century Britain." PhD diss., University of Cambridge, 2013.

———. "What Is Race? UNESCO, Mass Communication and Human Genetics in the Early 1950s." *History of Human Sciences* 28, no. 5 (2015): 80–107.

———. "Writing, Printing, Speaking: Rhesus Blood-Group Genetics and Nomenclatures in the Mid-Twentieth Century." *British Journal for the History of Science* 47, no. 2 (2014): 335–61.

Bangham, Jenny, and Soraya de Chadarevian. "Human Heredity after 1945: Moving Populations Centre Stage." *Studies in History and Philosophy of Biological and Biomedical Sciences* 47A (2014): 45–49.

———, eds. "Special Section—Heredity and the Study of Human Populations after 1945." *Studies in History and Philosophy of Biological and Biomedical Sciences* 47A (2014): 45–190.

Barahona, Ana. "Medical Genetics in Mexico: The Origins of Cytogenetics and the Health Care System." *Historical Studies in the Natural Sciences* 45 (2015): 147–73.

Barahona, Ana, Susana Pinar, and Francisco Ayala. "Introduction and Institutionalization of Genetics in Mexico." *Journal of the History of Biology* 38 (2005): 273–99.

Barnicot, N. A., and H. E. Huxley. "The Electron Microscopy of Unsectioned Human Chromosomes." *Annals of Human Genetics* 25 (1961): 253–58.

Barnicot, N. A., and P. J. Travers. "Comparison of the Human Karyotype in Various Populations." In *Proceedings of the Second International Congress of Human Genetics, Rome, September 6–12, 1961*, vol. 2, 1164–66. Rome: Istituto G. Mendel, 1963.

Barr, Murray L. "Cytological Tests of Sex" (letter to the editor). *Lancet* 267, no. 6906 (7 January 1956): 47.

——. "Human Cytogenetics: Some Reminiscences." *BioEssays* 9 (1988): 79–82.

Barr, Murray L., and Ewart G. Bertram. "A Morphological Distinction between Neurones of the Male and Female, and the Behaviour of the Nuclear Satellite during Accelerated Nucleoprotein Synthesis." *Nature* 163 (1949): 676–77.

Bashford, Alison. "Epilogue: Where Did Eugenics Go?" In *The Oxford Handbook of the History of Eugenics*, edited by Alison Bashford and Philippa Levine, 539–58. Oxford: Oxford University Press, 2010.

——. *Global Population: History, Geo-Politics and Life on Earth*. New York: Columbia University Press, 2014.

——. "Population, Geopolitics, and International Organizations in the Mid-Twentieth Century." *Journal of World History* 19 (2008): 327–47.

Bashford, Alison, and Philippa Levine, eds. *The Oxford Handbook of the History of Eugenics*. New York: Oxford University Press, 2010.

Bateman, Derek. "Secret Tests on 14 Children with 'Criminal Chromosomes.'" *Glasgow Herald*, 14 June 1979, 5.

Bauer, Dale R. "A Letter from the Publisher." *Medical World News*, 22 November 1968, 19.

Bauer, Susanne. "Mutations in Soviet Public Health Science: Post-Lysenko Medical Genetics, 1969–1991." *Studies in History and Philosophy of Biological and Biomedical Sciences* 47A (2014): 163–72.

Beatty, John. "Genetics in the Atomic Age: The Atomic Bomb Casualty Commission, 1947–1956." In *The Expansion of American Biology*, edited by K. B. Benson, J. Maienschein, and R. Rainger, 284–324. New Brunswick, NJ: Rutgers University Press, 1991.

——. "Scientific Collaboration, Internationalism, and Diplomacy: The Case of the Atomic Bomb Casualty Commission." *Journal of the History of Biology* 26 (1993): 205–31.

Beckwith, Jon. *Making Genes, Making Waves: A Social Activist in Science*. Cambridge, MA: Harvard University Press, 2002.

——. "Who Was Wronged in XYY?" *Hastings Center Report* 11, no. 2 (April 1981): 45.

Beckwith, Jon, Dirk Elseviers, Luigi Gorini, Chuck Mandansky, and Leslie Csonka. "Harvard XYY Study (Letter)." *Science* 187 (1975): 298.

Beckwith, Jon, and Jonathan King. "The XYY Syndrome: A Dangerous Myth." *New Scientist* 64 (14 November 1974): 474–76.

Beecher, Henry K. "Ethics and Clinical Research." *New England Journal of Medicine* 274, no. 24 (16 June 1966): 367–72.

Bell, Julia, and John Burdon Sanderson Haldane. "Blindness and Haemophilia in Man." *Proceedings of the Royal Society of London B* 123 (1937): 119–50.

Benirschke, Kurt, J. H. Edwards, A. Gropp, J. de Grouchy, R. T. Hill, R. L. Kirk, H. P. Klinger, O. J. Miller, P. E. Polani, A. Prokofjeva-Belgovskaja, M. Sasaki, C. C. Standley, and J. Steffen. "Standardization of Procedures for Chromosome Studies in Abortion." *Bulletin of the World Health Organization* 34 (1966): 765–82.

Benirschke, Kurt, and T. C. Hsu. *An Atlas of Mammalian Chromosomes.* Berlin: Springer, 1967–1977.

Benjamin, Ruha. "A Lab of Their Own: Genomic Sovereignty as Postcolonial Science Policy." *Policy and Society* 28 (2009): 341–55.

Berg, Kåre. "Introduction: Human Gene Mapping 6: Oslo Conference (1981), Sixth International Workshop on Human Gene Mapping." *Cytogenetics and Cell Genetics* 32 (1982): 1–3.

Bergsma, Daniel, ed. *ISCN 1978: An International System for Human Cytogenetic Nomenclature (1978): Report of the Standing Committee on Human Cytogenetic Nomenclature, Stockholm 1977.* Birth Defects: Original Article Series, vol. 14, no. 8. New York: National Foundation, March of Dimes, 1978.

Bergsma, Daniel, John L. Hamerton, and Harold P. Klinger, eds. "Chicago Conference: Standardization in Human Cytogenetics." *Birth Defects: Original Article Series* 2, no. 2 (December 1966): 1–21.

Beurton, Peter J., Raphael Falk, and Hans-Jörg Rheinberger, eds. *The Concept of the Gene in Development and Evolution: Historical and Epistemological Perspectives.* Cambridge: Cambridge University Press, 2000.

Bloom, Arthur D., James V. Neel, Kyoo W. Choi, Shozo Iida, and Napoleon Chagnon. "Chromosome Aberrations among the Yanomama Indians." *Proceedings of the National Academy of Sciences* 66 (1970): 920–27.

Bloom, A. D., S. Neriishi, N. Kamada, and T. Iseki. "Leukocyte Chromosome Studies of Adult and in-Utero Exposed Survivors of Hiroshima and Nagasaki." In *Human Radiation Cytogenetics: Proceedings of an International Symposium Held at Edinburgh, 12–15 October 1966*, edited by H. J. Evans, W. M. Court Brown, and A. S. McLean, 136–43. Amsterdam: North-Holland Publishing, 1967.

Bloom, A. D., S. Neriishi, N. Kamada, T. Iseki, and R. J. Keehn. "Cytogenetic Investigation of Survivors of the Atomic Bombings of Hiroshima and Nagasaki." *Lancet* 288, no. 7465 (1966): 672–74.

Botstein, D., R. White, M. Skolnick, and R. Davis. "Construction of a Genetic Linkage Map in Man Using Restriction Fragment Length Polymorphisms." *American Journal of Human Genetics* 32, no. 3 (1980): 314–31.

Bowker, Geoffrey C., and Susan Leigh Star. *Sorting Things Out: Classification and Its Consequences*. Cambridge, MA: MIT Press, 1999.

Bridges, Calvin B. "Sex in Relation to Chromosomes and Genes." *American Naturalist* 59, no. 661 (1925): 127–37.

Buckton, K. E., M. L. O'Riordan, P. A. Jacobs, J. A. Robinson, R. Hill, and H. J. Evans. "C- and Q-Band Polymorphisms in the Chromosomes of Three Human Populations." *Annals of Human Genetics* 46 (1976): 99–112.

Bunge, Raymond G. "Sex and the Olympic Games." *Journal of the American Medical Association* 173 (1960): 196.

———. "Sex and the Olympic Games No. 2." *Journal of the American Medical Association* 200, no. 10 (1967): 267.

Buzzati-Traverso, Adriano. "La materia e la vita: La superfemmina ha un X in più." *L'Espresso*, 18 October 1959, 8.

———. "Tutti da rifare i testi di genetica e di biologia: I segreti del sesso nel cromosoma ipsilon." *Il Giorno*, 22 July 1959, 9.

Cambrosio, Alberto, and Peter Keating. *Exquisite Specificity: The Monoclonal Antibody Revolution*. New York: Oxford University Press, 1995.

Campos, Luis. *Radium and the Secret of Life*. Chicago: University of Chicago Press, 2015.

Canguilhem, Georges. *On the Normal and the Pathological*. Dordrecht, The Netherlands: Reidel Publishing, 1978.

Capocci, Mauro, and Gilberto Corbellini. "Adriano Buzzati-Traverso and the Foundation of the International Laboratory of Genetics and Biophysics in Naples (1962–1969)." *Studies in History and Philosophy of Biological and Biomedical Sciences* 33 (2002): 489–513.

Carlson, Elof Axel. *Mendel's Legacy: The Origin of Classical Genetics*. Cold Spring Harbor, NY: Cold Spring Harbor Laboratory Press, 2004.

Carson, Rachel. *Silent Spring*. Boston: Houghton Mifflin, 1962.

Carter, K. Codell. "Early Conjectures That Down Syndrome Is Caused by Chromosomal Nondisjunction." *Bulletin of the History of Medicine* 3, no. 3 (2002): 528–63.

Cartwright, Lisa. *Screening the Body: Tracing Medicine's Visual Culture*. Minneapolis: University of Minnesota Press, 1995.

Casey, M. D., C. E. Blank, D. R. K. Street, L. J. Segall, J. H. McDougall, P.J. McGrath, and J. L. Skinner. "YY Chromosomes and Antisocial Behaviour." *Lancet* 288, no. 7468 (15 October 1966): 859–60.

Casey, M. D., L. J. Segall, D. R. K. Street, and C. E. Blank. "Sex Chromosome Abnormalities in Two State Hospitals for Patients Requiring Special Security." *Nature* 209 (1966): 641–42.

Caspari, Rachel. "From Types to Populations: A Century of Race, Physical Anthropology, and the American Anthropological Association." *American Anthropologist* 105, no. 1 (2003): 65–76.

Caspersson, Torbjörn, Gösta Lomakka, and Lore Zech. "The 24 Fluorescence Patterns of the Human Metaphase Chromosomes—Distinguishing Characters and Variability." *Hereditas* 67 (1971): 89–102.

Cavalli-Sforza, L. L., A. C. Wilson, C. R. Cantor, R. M. Cook-Deegan, and M.-C. King. "Call for a Worldwide Survey of Human Diversity: A Vanishing Opportunity for the Human Genome Project." *Genomics* 11 (1991): 490–91.

Cavalli-Sforza, L. L., L. A. Zonta, F. Nuzzo, L. Bernini, W. W. W. de Jong, P. Meera Khan, A. K. Ray, L. N. Went, M. Siniscalco, L. E. Nijenhuis, E. van Loghem, and G. Modiano. "Studies on African Pygmies: I. A Pilot Investigation of Babinga Pygmies in the Central African Republic (with an Analysis of Genetic Distances)." *American Journal of Human Genetics* 21 (1969): 252–74.

Chandra, H. Sharat, and David A. Hungerford. "Chromosome Studies of Todas of Southern India." *Human Biology* 38 (1966): 194–98.

Christie, D. A., and E. M. Tansey, eds. *Leukaemia: The Transcript of a Witness Seminar Held at the Wellcome Trust Centre for the History of Medicine at UCL, London, on 15 May 2001.* Wellcome Witnesses to Twentieth Century Medicine, vol. 15. London: Wellcome Trust Centre for the History of Medicine at UCL, 2003.

Christitch, Kosta. "Daniel Hugon est condamné à sept ans de réclusion criminelle pour l'assassinat d'une prostituée." *Le Monde*, 16 October 1968, 8.

"The Chromosomes of Man" (editorial). *Lancet* 273, no. 7075 (1959): 715–16.

Chu, Ernest H. Y. "Early Days of Mammalian Somatic Cell Genetics: The Beginning of Experimental Mutagenesis." *Mutation Research* 566 (2004): 1–8.

Clarke, A. E., and J. H. Fujimura, eds. *The Right Tools for the Job: At Work in Twentieth-Century Life Sciences.* Princeton, NJ: Princeton University Press, 1992.

Cole, Cheryl L. "One Chromosome Too Many?" In *The Olympics at the Millennium: Power, Politics, and the Games,* edited by Kay Schaffer and Sidonie Smith, 128–46. New Brunswick, NJ: Rutgers University Press, 2000.

Coleman, William. *Biology in the Nineteenth Century: Problems of Form, Function, and Transformation.* Cambridge: Cambridge University Press, 1985.

Collins, Kenneth John, and Joseph Sidney Weiner. *Human Adaptability: A History and Compendium of Research in the International Biological Programme.* London: Taylor & Francis, 1977.

Collyer, Susan, and Rhona De Mey. "Public Records and Recognition of Genetic Disease in Scotland." *Clinical Genetics* 31 (1987): 125–31.

Comfort, Nathaniel. "The Prisoner as Model Organism: Malaria Research at Stateville Penitentiary." *Studies in History and Philosophy of Biological and Biomedical Sciences* 40 (2009): 190–203.

——. *The Science of Human Perfection: How Genes Became the Heart of American Medicine*. New Haven, CT: Yale University Press, 2012.

Cottebrune, Anne. *Der planbare Mensch: Die Deutsche Forschungsgemeinschaft und die menschliche Vererbungswissenschaft (1920–1970)*. Stuttgart, Germany: Franz Steiner Verlag, 2008.

Court Brown, W. Michael. "Genetics and Crime." *Journal of the Royal College of Physicians* 1, no. 3 (April 1967): 311–18.

——. "Males with an XYY Sex Chromosome Complement." *Journal of Medical Genetics* 5 (1968): 341–59.

——. "Sex Chromosome Aneuploidy in Man and Its Frequency, with Special Reference to Mental Subnormality and Criminal Behavior." *International Review of Experimental Pathology* 7 (1968): 31–97.

——. "Sex Chromosomes and the Law." *Lancet* 280, no. 7254 (8 September 1962): 508–9.

——. "The Study of Human Sex Chromosome Abnormalities with Particular Reference to Intelligence and Behaviour." *Advancement of Science* 24 (1968): 390–97.

Court Brown, W. Michael, and John D. Abbatt. "The Incidence of Leukaemia in Ankylosing Spondylitis Treated with X-Rays: A Preliminary Report." *Lancet* 265, no. 6878 (1955): 1283–85.

Court Brown, W. Michael, and Richard Doll. "Appendix A: The Incidence of Leukaemia among the Survivors of the Atomic Bomb Explosions at Hiroshima and Nagasaki." In *The Hazards to Man of Nuclear and Allied Radiations*, Medical Research Council, 84–86. London: Her Majesty's Stationery Office, 1956. Cmnd. 9780.

——. "Appendix B: Leukaemia and Aplastic Anaemia in Patients Treated with X-Rays for Ankylosing Spondylitis." In *The Hazards to Man of Nuclear and Allied Radiations*, Medical Research Council, 87–89. London: Her Majesty's Stationery Office, 1956. Cmnd. 9780.

——. *Leukaemia and Aplastic Anaemia in Patients Irradiated for Ankylosing Spondylitis*. Medical Research Council Special Report Series 295. London: Her Majesty's Stationery Office, 1957.

Court Brown, W. M., D. G. Harnden, N. MacLean, and D. J. Mantle. *Abnormalities of the Sex Chromosome Complement in Man*. Privy Council Medical Research Council Special Report Series 305. London: Her Majesty's Stationery Office, 1964.

Couturier-Turpin, Marie-Hélène. "La découverte de la trisomie 21." *La revue du practicien* 55 (2005): 1385–89.

Coventry, Peter A., and John V. Pickstone. "From What and Why Did Genetics Emerge as a Medical Specialism in the 1970s in the UK? A Case History of Research, Policy and Services in the Manchester Region of the NHS." *Social Science and Medicine* 49 (1999): 1227–38.

Cowan, Ruth Schwartz. "Aspects of the History of Prenatal Diagnosis." *Fetal Diagnosis and Therapy* 8, supplement 1 (1993): 10–17.

———. *Heredity and Hope: The Case for Genetic Screening.* Cambridge, MA: Harvard University Press, 2008.

Crabtree, J. Michael, and Kenneth E. Moyer. *Bibliography of Aggressive Behavior: A Reader's Guide to the Research Literature.* New York: Alan R. Liss, 1977.

Creager, Angela N. H. *Life Atomic: A History of Radioisotopes in Science and Medicine.* Chicago: University of Chicago Press, 2013.

———. "Mutation in the Atomic Age." *Historical Studies in the Natural Sciences* 45 (2015): 14–48.

———. "Phosphorus-32 in the Phage Group: Radioisotopes as Historical Tracers of Molecular Biology." *Studies in History and Philosophy of Biological and Biomedical Sciences* 40 (2009): 29–42.

Cremer, T., and C. Cremer. "Chromosome Territories, Nuclear Architecture and Gene Regulation in Mammalian Cells." *Nature Reviews Genetics* 2 (2001): 292–301.

Crick, Francis H. C. "The Double Helix: A Personal View." *Nature* 248 (1974): 766–69.

———. "General Model for the Chromosomes of Higher Organisms." *Nature* 234 (1971): 25–27.

———. *Of Molecules and Men.* Seattle: University of Washington Press, 1966.

———. "Postscript." In *Chromosomes Today*, vol. 6, *Proceedings of the Sixth International Chromosome Conference Held in Helsinki, Finland, August 29–31, 1977*, edited by A. de la Chapelle and M. Sorsa, 403–6. Amsterdam: Elsevier/North-Holland Biomedical Press, 1977.

———. *What Mad Pursuit: A Personal View of Scientific Discovery.* Harmondsworth, UK: Penguin Books, 1990.

"Criminal Behaviour and the Y Chromosome" (editorial). *British Medical Journal* 1, no. 5532 (14 January 1967): 64–65.

Cryle, Peter, and Elizabeth Stephens. *Normality: A Critical Genealogy.* Chicago: University of Chicago Press, 2017.

Culliton, Barbara J. "Patients' Rights: Harvard Is Site of Battle over X and Y Chromosomes." *Science* 186 (22 November 1974): 715–17.

Curry, Helen Anne. *Evolution Made to Order: Plant Breeding and Technological Innovation in Twentieth Century America.* Chicago: University of Chicago Press, 2016.

Dahlberg, Gunnar. *Race, Reason and Rubbish: A Primer of Race Biology.* Translated by Lancelot Thomas Hogben. London: Allen and Unwin, 1942.

Darby, Sarah. "A Conversation with Sir Richard Doll." *Epidemiology* 14 (2003): 375–79.

Darlington, Cyril D. "The Chromosomes as We See Them." In *Chromosomes Today*, vol. 1, edited by C. D. Darlington and K. R. Lewis, 1–6. Edinburgh: Oliver and Boyd, 1966.

Darlington, Cyril D., and E. K. Janaki Ammal. *Chromosome Atlas of Cultivated Plants*. London: Allen and Unwin, 1945.

Daston, Lorraine. "On Scientific Observation." *Isis* 99 (2008): 97–110.

——, ed. *Things That Talk: Object Lessons from Art and Science*. New York: Zone Books, 2004.

Daston, Lorraine, and Peter Galison. *Objectivity*. New York: Zone Books, 2007.

Daston, Lorraine, and Elizabeth Lunbeck. "Introduction: Observation Observed." In *Histories of Scientific Observation*, edited by Lorraine Daston and Elizabeth Lunbeck, 1–9. Chicago: University of Chicago Press, 2011.

Davenport, Charles B. "Mendelism in Man." In *Proceedings of the Sixth International Congress of Genetics, Ithaca, New York 1932*, vol. 1, *Transactions and General Addresses*, edited by Donald F. Jones, 135–40. Menasha, WI: Brooklyn Botanic Garden, 1932.

Davidson, William M., and David Robertson Smith, eds. *Proceedings of the Conference on Human Chromosomal Abnormalities*. London: Staples Press, 1961.

Davies, Chris. *Changing Society: A Personal History of Scope (Formerly the Spastics Society), 1952–2000*. London: Scope, 2002.

Davis, F. James. *Who Is Black? One Nation's Definition*. University Park: Pennsylvania State University Press, 2001.

Davis, Harry M. "We Enter a New Era—The Atomic Age." *New York Times*, 12 August 1945, SM3.

De Bont, Raf. "'Primitives' and Protected Areas: International Conservation and the 'Naturalization' of Indigenous People, ca. 1910–1975." *Journal of the History of Ideas* 76 (2015): 215–36.

de Chadarevian, Soraya. "Chromosome Photography and the Human Karyotype." *Historical Studies in the Natural Sciences* 45, no. 1 (2015): 115–46.

——. *Designs for Life: Molecular Biology after World War II*. Cambridge: Cambridge University Press, 2002.

——. "Human Population Studies and the World Health Organization." *Dynamis* 35 (2015): 359–88.

——. "The Making of an Entrepreneurial Science: Biotechnology in Britain, 1975–1995." *Isis* 102 (2011): 601–33.

——. "Mapping Development or How Molecular Is Molecular Biology?" *History and Philosophy of the Life Sciences* 22 (2000): 335–50.

——. "Mice and the Reactor: The 'Genetic Experiment' in 1950s Britain." *Journal of the History of Biology* 39 (2006): 707–35.

de Chadarevian, Soraya, and Harmke Kamminga, eds. Introduction to *Molecularizing Biology and Medicine: New Practices and Alliances, 1910s–1970s*,

edited by Soraya de Chadarevian and Harmke Kamminga, 1–16. Amsterdam: Harwood Academic Publishers, 1984.

de Chadarevian, Soraya, and Theodore Porter, eds. "Histories of Data and the
Database." Special issue of *Historical Studies in the Natural Sciences* 48,
no. 5 (2018).

Deepe Keever, Beverly A. *News Zero: The New York Times and the Bomb.*
Monroe, ME: Common Courage Press, 2004.

de Garay, Alfonso L., Louis Levine, and J. E. Lindsay Carter, eds. *Genetic and
Anthropological Studies of Olympic Athletes.* New York: Academic Press,
1974.

de Grouchy, Jean, and Catherine Turleau. *Clinical Atlas of Human Chromosomes.* New York: John Wiley & Sons, 1977.

de la Chapelle, Albert. "The Use and Misuse of Sex Chromatin Screening for
'Gender Identification' of Female Athletes." *Journal of the American Medical Association* 256 (1986): 1920–23.

Dent, Rosanna. "Studying Indigenous Brazil: The Xavante and the Human Sciences, 1958–2015." PhD diss., University of Pennsylvania, 2017.

De Sio, Fabio. "Genetics and International Cooperation: Giuseppe Montalenti
and Modern Biology." *Medicina nei secoli* 18 (2006): 121–58.

Dolphin, G. W., and D. C. Lloyd. "The Significance of Radiation-Induced Chromosome Abnormalities in Radiobiological Protection." *Journal of Medical
Genetics* 11 (1974): 181–89.

Donahue, Roger P., Wilma B. Bias, James H. Renwick, and Victor A. McKusick.
"Probable Assignment of the Duffy Blood Group Locus to Chromosome 1
in Man." *Proceedings of the National Academy of Sciences* 61, no. 3 (1968):
949–55.

Down, J. Langdon H. "Observations on an Ethnic Classification of Idiots." *Clinical Lecture Reports* 3 (1866): 259–62.

Drucker, Brian J. "Janet Rowley (1925–2013)." *Nature* 505 (2014): 484.

Durant, John R. "The Beast in Man: An Historical Perspective on the Biology
of Human Aggression." In *The Biology of Aggression*, edited by P. Brain and
D. Benton, 17–47. Leiden, The Netherlands: Sijthoff, 1981.

Duster, Troy. *Backdoor to Eugenics.* London: Routledge, 1990.

Dutrillaux, B., J. Couturier, Claude-Lise Richer, and Evani Viegas-Péquignot.
"Sequence of DNA Replication in 277 R- and Q-Bands of Human Chromosomes Using BrdU Treatment." *Chromosoma* 58 (1976): 51–61.

Eder, Sandra. "Gender and Cortisone: Clinical Practice and Transatlantic Exchange in the Medical Management of Intersex in the 1950s." *Bulletin of the
History of Medicine* 92 (2018): 604–33.

———. "The Volatility of Sex: Intersexuality, Gender and Clinical Practice in the
1950s." *Gender & History* 22 (2010): 692–707.

Edwards, Anthony W. F. "Linkage Methods in Human Genetics before the Computer." *Human Genetics* 118 (2005): 515–30.

Edwards, J. H. "Painless Skin Biopsy." *Lancet* 275, no. 7122 (27 February 1960): 496.

Ellerström, Sven, and Arne Hagberg. "The Cyto-Genetic Department of the Swedish Seed Association 1931–1961. A Brief Survey of the Work Carried Out during the Passed [*sic*] 30 Years." *Sveriges Utsädesförenings Tidskrift* 72 (1962): 192–209.

Elzen, Boelie. "Two Ultracentrifuges: A Comparative Study of the Social Construction of Artefacts." *Social Studies of Science* 16 (1986): 621–62.

Engel, Eric. "Guest Editorial: The Making of an XYY." *American Journal of Mental Deficiency* 77, no. 2 (1972): 123–27.

Epstein, Steven. *Inclusion: The Politics of Difference in Medical Research*. Chicago: University of Chicago Press, 2007.

"Errors in the Factory—1." *World Health: The Magazine of the World Health Organization* (September 1966), 12–17.

Escoffier-Lambiotte, Claudine. "L'hérédité de la violence." *Le Monde*, 16 October 1968, 8.

Evans, H. J., K. E. Buckton, G. E. Hamilton, and A. Carothers. "Radiation-Induced Chromosome Aberrations in Nuclear Dockyard Workers." *Nature* 227 (1979): 531–34.

Evans, H. J., W. M. Court Brown, and A. S. McLean, eds. *Human Radiation Cytogenetics*. Amsterdam: North-Holland Publishing, 1967.

Evans, Jane A., John L. Hamerton, and Arthur Robinson, eds. *Children and Young Adults with Chromosome Aneuploidy: Follow-up, Clinical, and Molecular Studies. Proceedings of the 5th International Workshop on Sex Chromosome Aneuploidy Held at Minaki, Ontario, Canada, June 7–10, 1989*. Birth Defects: Original Article Series. New York: Wiley-Liss, 1991.

"Faculty Debates Chromosome Study." *Harvard Medical Alumni Bulletin* 49, no. 5 (May–June 1975): 4–5.

Fausto-Sterling, Anne. *Sexing the Body: Gender Politics and the Construction of Sexuality*. New York: Basic Books, 2000.

Ferguson-Smith, Malcolm A. "Cytogenetics and Early Days at the Moore Clinic with Victor McKusick." In *Victor McKusick and the History of Medical Genetics*, edited by Krishna R. Dronamraju and Clair A. Francomano, 53–66. New York: Springer, 2012.

———. "From Chromosome Number to Chromosome Map: The Contribution of Human Cytogenetics to Genome Mapping." In *Chromosomes Today*, edited by A. T. Sumner and A. C. Chandley, 11:3–19. London: Chapman & Hall, 1993.

———. "Gender Verification and the Place of XY Females in Sport." In *Oxford*

Textbook of Sports Medicine, edited by M. Harries, C. Williams, W. D. Stanish, and L. J. Micheli, 355–65. Oxford: Oxford University Press, 1998.

——. "Putting Medical Genetics into Practice." *Annual Review of Genomics and Human Genetics* 12 (2011): 1–23.

——. "Sex Chromatin, Klinefelter's Syndrome and Mental Deficiency." In *The Sex Chromatin*, edited by Keith L. Moore, 277–315. Philadelphia: Saunders Company, 1966.

Finch, John. *A Nobel Fellow on Every Floor: A History of LMB.* Cambridge, UK: MRC Laboratory of Molecular Biology, 2008.

Fleck, Ludwig. *Genesis and Development of a Scientific Fact.* Chicago: University of Chicago Press, 1979.

Ford, C. E. "Human Cytogenetics: Its Present Place and Future Possibilities." *American Journal of Human Genetics* 1960, no. 12 (1960): 104–17.

Ford, C. E., P. A. Jacobs, and L. G. Lajtha. "Human Somatic Chromosomes." *Nature* 181 (1958): 1565–68.

Ford, C. E., K. W. Jones, O. J. Miller, Ursula Mittwoch, S. L. Penrose, M. Ridler, and A. Shapiro. "The Chromosomes in a Patient Showing Both Mongolism and the Klinefelter Syndrome." *Lancet* 273, no. 7075 (1959): 709–10.

Ford, Charles E., Kenneth W. Jones, Paul E. Polani, José Carlos Cabral de Almeida, and John H. Briggs. "A Sex-Chromosome Anomaly in a Case of Gonadal Dysgenesis (Turner's Syndrome)." *Lancet* 273, no. 7075 (4 April 1959): 711–13.

Ford, E. H. R. *Human Chromosomes.* New York: Academic Press, 1973.

Ford, E. H. R., K. Thurley, and D. H. M. Woollam. "Electron-Microscopic Observations on Whole Mitotic Chromosomes." *Journal of Anatomy* 103 (1968): 143–50.

Fortun, Michael. "Projecting Speed Genomics." In *The Practices of Human Genetics*, edited by Michael Fortun and Everett Mendelsohn, 21:25–48. Sociology of the Sciences Yearbook 21. Dordrecht, The Netherlands: Kluwer, 1999.

Fox, Richard G. "XYY Offender: A Modern Myth." *Journal of Criminal Law and Criminology* 62, no. 1 (1971): 59–73.

Fraccaro, Marco, and Jan Lindsten. "Le malattie cromosomiche: Un nuovo ramo della medicina." *Sapere* 640 (April 1963): 197–204.

Franzke, Alice W. "Telling Parents about XYY Sons." *New England Journal of Medicine* 293 (10 July 1975): 100–101.

Fuchs, F., E. Freiesleben, E. E. Knudsen, and P. Riis. "Antenatal Detection of Hereditary Diseases." *Acta Genetica et Statistica Medica* 6 (1956): 261–63.

Fuchs, F., and P. Riis. "Antenatal Sex Determination." *Nature* 177 (1956): 330.

Galison, Peter. *Image and Logic: A Material Culture of Microphysics.* Chicago: University of Chicago Press, 1997.

Gannett, Lisa, and James Griesemer. "The ABO Blood Groups: Mapping the History and Geography of Genes in Homo Sapiens." In *Classical Genetic Re-*

search and Its Legacy: The Mapping Cultures of Twentieth-Century Genetics, edited by Hans-Jörg Rheinberger and Jean-Paul Gaudillière, 119–72. London: Routledge, 2004.

García-Sancho, Miguel. "The Proactive Historian: Methodological Opportunities around the New Archival Evidence on Modern Genomics." *Studies in History and Philosophy of Biological and Biomedical Sciences* 55 (2016): 70–82.

Gaudillière, Jean-Paul. "Le syndrome nataliste: Étude de l'hérédité, pédiatrie et eugénisme en France (1920–1960)." *Histoire de la médecine et des sciences* 13 (1997): 1165–71.

———. "Whose Work Shall We Trust? Geneticists, Pediatrics, and Hereditary Diseases in Postwar France." In *Controlling Our Destinies: Historical, Philosophical, Ethical, and Theological Perspectives on the Human Genome Project*, edited by Phillip R. Sloan, 17–46. Notre Dame, IN: University of Notre Dame Press, 2000.

Gaudillière, Jean-Paul, and Hans-Jörg Rheinberger, eds. *From Molecular Genetics to Genomics: The Mapping Cultures of Twentieth-Century Genetics*. London: Routledge, 2004.

Gausemeier, Bernd, Staffan Müller-Wille, and Edmund Ramsden, eds. *Human Heredity in the Twentieth Century*. London: Pickering and Chatto, 2013.

Gautier, Marthe, and Peter Harper. "Fiftieth Anniversary of Trisomy 21: Returning to a Discovery." *Human Genetics* 126 (2009): 317–24.

Gilbert, Walter. "A Vision of the Grail." In *The Code of Codes: Scientific and Social Issues in the Human Genome Project*, edited by Daniel J. Kevles and Leroy Hood, 83–97. Cambridge, MA: Harvard University Press, 1992.

Glass, Bentley. "Less Than Golden Future" (reply to letters by B. Raymond Fink and Rudolph Steinberger). *Science* 172 (9 April 1971): 111–12.

———. "Science: Endless Horizon or Golden Age?" *Science* 171 (8 January 1971): 23–29.

———. "Theophilus Shickel Painter, 1889–1969." In *National Academy of Sciences Biographical Memoirs*, 59:308–37. Washington, DC: National Academy of Sciences, 1990.

———. "What Price the Perfect Baby?" (reply to letter by Leon R. Kass). *Science* 173 (9 July 1971): 103–4.

Gordin, Michael D. *Five Days in August: How World War II Became a Nuclear War*. Princeton, NJ: Princeton University Press, 2007.

———. "Lysenko Unemployed: Soviet Genetics after the Aftermath." *Isis* 109 (2018): 56–78.

Gould, Stephen Jay. *The Mismeasure of Man*. New York: W. W. Norton, 1996.

Green, Jeremy. "Media Sensationalisation and Science: The Case of the Criminal Chromosome." In *Expository Science: Forms and Functions of Popularisation*, edited by Terry Shinn and Richard Whitley, 139–61. Boston: D. Reidel, 1985.

Greene, Gayle. *The Woman Who Knew Too Much: Alice Stewart and the Secrets of Radiation*. Ann Arbor: University of Michigan Press, 1999.

Gusella, David F., N. S. Wexler, P. M. Conneally, S. L. Naylor, M. A. Anderson, R. E. Tanzi, P. C. Watkins, K. Ottina, M. R. Wallace, A. Y. Sakaguchi, et al. "A Polymorphic DNA Marker Genetically Linked to Huntington Disease." *Nature* 306, no. 5940 (17–23 November 1983): 234–38.

Guyer, Michael F. "A Note on the Accessory Chromosomes of Man." *Science* 39 (15 May 1914): 721–22.

Ha, Nathan Q. "Diagnosing Sex Chromatin: A Binary for Every Cell." *Historical Studies in the Natural Sciences* 45 (2015): 49–84.

———. "Marking Bodies: A History of Genetic Sex in the Twentieth Century." PhD diss., Princeton University, 2011.

———. "The Riddle of Sex: Biological Theories of Sexual Difference in the Early Twentieth Century." *Journal of the History of Biology* 44 (2011): 505–46.

Hacker, Barton C. *The Dragon's Tail: Radiation Safety in the Manhattan Project, 1942–1946*. Berkeley: University of California Press, 1987.

———. *Elements of Controversy: The Atomic Energy Commission and Radiation Safety in Nuclear Weapons Testing, 1947–1974*. Berkeley: University of California Press, 1994.

Hacking, Ian. *Representing and Intervening: Introductory Topics in the Philosophy of Natural Science*. Cambridge: Cambridge University Press, 1983.

———. "Styles of Scientific Reasoning." In *Post-Analytic Philosophy*, edited by John Rajchman and Cornel West, 145–65. New York: Columbia University Press, 1985.

Haldane, John Burdon Sanderson. "The Formal Genetics of Man." *Proceedings of the Royal Society of London B* 153 (1948): 147–70.

———. "A Provisional Map of a Human Chromosome." *Nature* 137 (1936): 398–400.

Hall, M. Ann. *The Girl and the Game: A History of Women's Sport in Canada*. Peterborough, ON: Broadview Press, 2002.

Hamblin, Jacob Darwin. "'A Dispassionate and Objective Effort': Negotiating the First Study on the Biological Effects of Atomic Radiation." *Journal of the History of Biology* 40 (2007): 147–77.

Hamerton, John L. "Human Cytogenetic Registries." *Humangenetik* 29 (1975): 177–81.

———. *Human Cytogenetics*. 2 vols. New York: Academic Press, 1971.

———. "Human Population Cytogenetics: Dilemmas and Problems." *American Journal of Human Genetics* 28 (1976): 107–22.

Hamerton, J. L., M. Fraccaro, L. De Carli, F. Nuzzo, H. P. Klinger, L. Hulliger, A. Taylor, and E. M. Lang. "Somatic Chromosomes of the Gorilla." *Nature* 192, no. 4799 (1961): 225–28.

Hamerton, John L., Patricia A. Jacobs, and Harold P. Klinger, eds. "Paris Conference (1971): Standardization in Human Cytogenetics." *Cytogenetics* 11 (1972): 313–62.

Hamerton, John L., and Harold P. Klinger. *Paris Conference (1971), Supplement (1975).* Birth Defects: Original Article Series, vol. 11, no. 9. New York: National Foundation, March of Dimes, 1975.

Hamerton, J. L., H. P. Klinger, D. E. Mutton, and E. M. Lang. "The London Conference on the Normal Human Karyotype, 28th–30th August, 1963." *Cytogenetics* 2 (1963): 264–68.

Hamerton, John L., Angela I. Taylor, Roslyn Angell, and V. Mary McGuirre. "Chromosome Investigations of a Small Isolated Human Population: Chromosome Abnormalities and Distribution of Chromosome Counts According to Age and Sex among the Population of Tristan da Cunha." *Nature* 206 (1965): 1232–34.

Harman, Oren S. *The Man Who Invented the Chromosome: A Life of Cyril Darlington.* Cambridge, MA: Harvard University Press, 2004.

———. "Unformed Minds: Juveniles, Neuroscience, and the Law." *Studies in History and Philosophy of Biological and Biomedical Sciences* 44 (2013): 455–59.

Harnden, David G. "The Chromosomes." In *Recent Advances in Human Genetics*, edited by L. S. Penrose and Helen Lang Brown, 19–38. London: J. & A. Churchill, 1961.

———. "Early Studies on Human Chromosomes." *BioEssays* 18 (1996): 162–68.

———. "A Human Skin Culture Technique Used for Cytological Examination." *British Journal of Experimental Pathology* 41 (1960): 31–37.

Harper, Peter S. "The Discovery of the Human Chromosome Number in Lund, 1955–1956." *Human Genetics* 119 (2006): 226–32.

———. *First Years of Human Chromosomes: The Beginning of Human Cytogenetics.* Bloxham, UK: Scion Publishing, 2006.

———. "Paul Polani and the Development of Medical Genetics." *Human Genetics* 120 (2007): 723–31.

———. *A Short Story of Medical Genetics.* Oxford: Oxford University Press, 2008.

Harris, Henry. *The Cells of the Body: A History of Somatic Cell Genetics.* Cold Spring Harbor, NY: Cold Spring Harbor Laboratory Press, 1995.

Harris, Henry, J. F. Watkins, G. L. Campbell, E. P. Evans, and C. E. Ford. "Mitosis in Hybrid Cells Derived from Mouse and Man." *Nature* 207 (7 August 1965): 606–8.

Harrison, G. A., J. S. Weiner, J. M. Tanner, and N. A. Barnicot. *Human Biology: An Introduction to Human Evolution, Variation, Growth and Ecology. Second Edition.* Oxford: Oxford University Press, 1977.

Hastings Center. "Special Supplement: The XYY Controversy: Researching Violence and Genetics." *Hastings Center Report* 10, no. 4 (4 August 1980): 1–31.

Hauschka, Theodore S., John E. Hasson, Milton N. Goldstein, George F. Koepf, and Avery A. Sandberg. "An XYY Man with Progeny Indicating Familial Tendency to Non-Disjunction." *American Journal of Human Genetics* 14 (1962): 22–30.

Henare, Amira, Martin Holbraad, and Sari Wastell, eds. *Thinking through Things: Theorising Artefacts Ethnographically.* London: Routledge, 2007.

Herran, Néstor. "Isotope Networks: Training, Sales and Publications, 1946–1965." *Dynamis* 29 (2009): 285–306.

———. "Spreading Nucleonics: The Isotope School at the Atomic Energy Research Establishment, 1951–67." *British Journal for the History of Science* 39 (2006): 569–86.

Hilgartner, Stephen. "Constituting Large-Scale Biology: Building a Regime of Governance in the Early Years of the Human Genome Project." *BioSocieties* 8 (2013): 397–416.

———. *Reordering Life: Knowledge and Control in the Genomics Revolution.* Cambridge, MA: MIT Press, 2017.

Himsworth, Harold P. "Leukaemia and Ankylosing Spondylitis." *Lancet* 265, no. 6857 (29 January 1955): 250.

Hirschhorn, K., and S. Boyer. "Report of the Committee on In Situ Hybridization—New Haven Conference (1973): First International Workshop on Human Gene Mapping." *Cytogenetics and Cell Genetics* 13 (1974): 55–57.

Hogan, Andrew J. "Disrupting Genetic Dogma: Bridging Cytogenetics and Molecular Biology in Fragile X Research." *Historical Studies in the Natural Sciences* 45 (2015): 174–97.

———. *Life Histories of Genetic Disease: Patterns and Prevention in Postwar Medical Genetics.* Baltimore: Johns Hopkins University Press, 2016.

———. "The 'Morbid Anatomy' of the Human Genome: Tracing the Observational and Representational Approaches of Postwar Genetics and Biomedicine." *Medical History* 58 (2014): 315–36.

Hogarth, Stuart, Michael M. Hopkins, and Victor Rodriguez. "A Molecular Monopoly? HPV Testing, the Pap Smear and the Molecularisation of Cervical Cancer Screening in the USA." *Sociology of Health and Illness* 34 (2012): 234–50.

Hogben, Lancelot. *Genetic Principles in Medicine and Social Science.* London: Williams and Norgate, 1931.

Holmes, Frederic Lawrence. *Meselson, Stahl, and the Replication of DNA: A History of "The Most Beautiful Experiment in Biology."* New Haven, CT: Yale University Press, 2001.

———. "Seymour Benzer and the Convergence of Molecular Biology with Classical Genetics." In *From Molecular Genetics to Genomics: The Mapping Cultures of Twentieth-Century Genetics*, edited by Jean-Paul Gaudillière and Hans-Jörg Rheinberger, 42–62. London: Routledge, 2004.

———. "Seymour Benzer and the Definition of the Gene." In *The Concept of the*

Gene in Development and Evolution: Historical and Epistemological Perspectives, edited by R. Falk and H.-J. Rheinberger P. Beurton, 115–55. Cambridge: Cambridge University Press, 2000.

Holmes, Frederic Lawrence, and William C. Summers. *Reconceiving the Gene: Seymour Benzer's Adventures in Phage Genetics*. New Haven, CT: Yale University Press, 2006.

Hood-Williams, John. "Sexing the Athletes." *Sociology of Sports Journal* 12 (1995): 290–305.

Hook, Ernest B. "Behavioral Implications of the Human XYY Genotype." *Science* 179 (1973): 139–50.

———. "Extra Sex Chromosomes and Human Behavior: The Nature of the Evidence Regarding XYY, XXY, XXYY, and XXX Genotype." In *Genetic Mechanisms of Sexual Development*, edited by H. L. Vallet and I. H. Porter, 437–63. New York: Academic Press, 1979.

———. "Geneticophobia and the Implications of Screening for the XYY Genotype in Newborn Infants." In *Genetics and the Law*, edited by A. Milunsky and G. J. Annas, 73–86. New York: Plenum Publishing, 1995.

Hopkins, Michael M. "The Hidden Research System: The Evolution of Cytogenetic Testing in the National Health Service." *Science as Culture* 15 (2006): 253–76.

Hopwood, Nick. "Embryology." In *The Cambridge History of Science*, vol. 6, *The Modern Biological and Earth Sciences*, edited by Peter J. Bowler and John V. Pickstone, 285–315. Cambridge: Cambridge University Press, 2009.

———. *Haeckel's Embryos: Images, Evolution, and Fraud*. Chicago: University of Chicago Press, 2015.

———. "Visual Standards and Disciplinary Change: Normal Plates, Tables and Stages in Vertebrate Embryology." *History of Science* 43 (2005): 239–303.

Hsu, T. C. *Human and Mammalian Cytogenetics: An Historical Perspective*. New York: Springer, 1979.

———. "Mammalian Chromosomes in Vitro: I. The Karyotype of Man." *Journal of Heredity* 43 (1952): 167–72.

———. "My Favorite Cytological Subject: Chromosomes." *BioEssays* 14 (1992): 785–89.

Hubbard, Ruth. "Abortion and Disability: Who Should and Should Not Inhabit the World?" In *The Disability Studies Reader*, edited by Lennard J. Davis, 74–86. New York: Routledge, 2013.

Human Chromosomes Study Group. "A Proposed Standard of Nomenclature of Human Mitotic Chromosomes." *Cerebral Palsy Bulletin*, supplement 2, no. 3 (1960): 1–9.

Human Cytogenetic Registers: A Description of Nine Systems with Some Recommendations. Birth Defects: Original Article Series, vol. 13, no. 4. New York: National Foundation, March of Dimes, 1977.

Hungerford, David A. "Some Early Studies of Human Chromosomes, 1879–1955." *Cytogenetics and Cell Genetics* 20 (1978): 1–11.

Hungerford, David A., Eugene Giles, and Charlotte G. Creech. "Chromosome Studies of Eastern New Guinea Natives." *Current Anthropology* 6 (1965): 109–10.

Hungerford, D. A., S. Makino, M. Sasaki, A. A. Awa, and Gloria B. Balaban. "Chromosome Studies of the Ainu Population of Hokkaido." *Cytogenetics* 8 (1969): 74–79.

Huxley, Julian. "Biological Basis of Psychosocial Evolution. Human Biology: An Introduction to Human Evolution, Variation and Growth by G. A. Harrison, J. S. Weiner, J. M. Tanner and N. A. Barnicot (Oxford: Clarendon Press 1964)" (review). *Nature* 204 (1964): 950–51.

Hyun, Jaehwan. "Making Postcolonial Connections: The Role of the Japanese Research Network in the Emergence of Human Genetics in South Korea, 1941–68." *Korean Journal for the History of Science* 39 (2017): 293–324.

Ingram, Vernon. "Gene Mutations in Human Haemoglobin: The Chemical Difference between Normal Human and Sickle-Cell Haemoglobin." *Nature* 180 (1957): 326–28.

———. "A Specific Chemical Difference between the Globins in Normal Human and Sickle-Cell Anaemia Haemoglobin." *Nature* 178 (1956): 792–94.

Irgens, L. M. "The Origin of Registry-Based Medical Research and Care." *Acta Neurologica Scandinavica* 126, supplement 195 (2012): 4–6.

Jacobs, Patricia A. "Counting and Analysis of Human Chromosomes." In *Mathematics and Computer Science in Biology and Medicine: Proceedings of a Conference Held by the Medical Research Council in Association with the Health Departments, Oxford, July 1964*, edited by Medical Research Council, 185–88. London: Her Majesty's Stationery Office, 1965.

———. "Human Chromosome Heteromorphisms (Variants)." *Progress in Medical Genetics*, n.s., 2 (1977): 251–74.

———. "An Opportune Life: 50 Years in Human Cytogenetics." *Annual Review of Genomics and Human Genetics* 15 (2014): 29–46.

———. "Population Surveillance: A Cytogenetic Approach." In *Genetic Epidemiology*, edited by Newton E. Morton and Chin Sik Chung, 463–81. New York: Academic Press, 1978.

———. "The William Allan Memorial Award Address: Human Population Cytogenetics: The First Twenty-Five Years." *American Journal of Human Genetics* 34 (1982): 689–98.

Jacobs, Patricia A., Albert G. Baikie, W. Michael Court Brown, Thomas N. MacGregor, and David G. Harnden. "Evidence for the Existence of the Human 'Super Female.'" *Lancet* 274, no. 7100 (26 September 1959): 423–25.

Jacobs, Patricia A., A. G. Baikie, W. M. Court Brown, and J. A. Strong. "The Somatic Chromosomes in Mongolism." *Lancet* 273, no. 7075 (4 April 1959): 710.

Jacobs, Patricia A., Muriel Brunton, Marie M. Melville, Robert P. Brittain, and William F. McClemont. "Aggressive Behaviour, Mental Sub-Normality and the XYY Male." *Nature* 208 (1965): 1351–52.

Jacobs, Patricia A., W. H. Price, W. M. Court Brown, R. P. Brittain, and P. B. Whatmore. "Chromosome Studies on Men in a Maximum Security Hospital." *Annals of Human Genetics* 31 (1968): 339–58.

Jacobs, Patricia A., W. H. Price, and Pamela Law, eds. *Human Population Cytogenetics.* Baltimore: Williams and Wilkins Company, 1970.

Jacobs, Patricia A., William H. Price, Shirley Richmond, and R. A. W. Ratcliff. "Chromosome Surveys in Penal Institutions and Approved Schools." *Journal of Medical Genetics* 8 (1971): 49–58.

Jacobs, Patricia A., and John A. Strong. "A Case of Human Intersexuality Having a Possible XXY Sex-Determining Mechanism." *Nature* 183 (31 January 1959): 302–3.

Jay, Martin. *Downcast Eyes: The Denigration of Vision in Twentieth-Century French Thought.* Berkeley: University of California Press, 1994.

Jensen, Arthur. "How Much Can We Boost IQ and School Achievement?" *Harvard Educational Review* 39 (1969): 1–123.

Jones, E. M., and E. M. Tansey, eds. *Human Gene Mapping Workshops c. 1973–c. 1991.* Wellcome Witnesses to Contemporary Medicine, vol. 54. London: Queen Mary University of London, 2015.

Jones, Matthew L. "How We Became Instrumentalists (Again): Data Positivism since World War II." *Historical Studies in the Natural Sciences* 48, no. 5 (2018): 673–84.

Jung, Monica de Paula, Maria Helena Cabral de Almeida Cardoso, Maria Auxiliadora Monteiro Villar, and Juan Clinton Lerena Jr. "Revisiting Establishment of the Etiology of Turner Syndrome." *História, ciências, saúde—Manguinhos* 16 (2009): 361–76.

Kay, Lily. *Who Wrote the Book of Life? A History of the Genetic Code.* Stanford, CA: Stanford University Press, 2000.

Keating, Peter, and Alberto Cambrosio. *Biomedical Platforms: Realigning the Normal and the Pathological in Late-Twentieth-Century Medicine.* Cambridge, MA: MIT Press, 2003.

———. *Cancer on Trial: Oncology as a New Style of Practice.* Chicago: University of Chicago Press, 2012.

———. "The New Genetics and Cancer: The Contributions of Clinical Medicine in the Era of Biomedicine." *Journal of the History of Medicine and Allied Sciences* 56 (2001): 321–52.

———. "'Ours Is an Engineering Approach': Flow Cytometry and the Constitution of Human T-Cell Subsets." *Journal of the History of Biology* 27, no. 3 (1994): 449–79.

Keirns, Carla. "Seeing Patterns: Models, Visual Evidence, and Pictorial Communication in the World of Barbara McClintock." *Journal of the History of Biology* 32 (1999): 163–96.

Keller, Evelyn Fox. *The Century of the Gene*. Cambridge, MA: Harvard University Press, 2000.

———. "The Postgenomic Genome." In *Postgenomics: Perspectives on Biology after the Gene*, edited by Sarah S. Richardson and Hallam Stevens, 9–31. Durham, NC: Duke University Press, 2013.

Kelty, Christopher M. "This Is Not an Article: Model Organism Newsletters and the Question of 'Open Science.'" *BioSocieties* 7, no. 2 (2012): 140–68.

Kevles, Daniel J. *In the Name of Eugenics: Genetics and the Uses of Human Heredity*. Cambridge, MA: Harvard University Press, 1995.

Kodani, Masuo. "The Supernumerary Chromosome of Man." *American Journal of Human Genetics* 10 (1958): 125–40.

———. "Three Chromosome Numbers in Whites and Japanese." *Science* 127 (1958): 1339–40.

———. "Three Diploid Chromosome Numbers in Man." *Proceedings of the National Academy of Sciences* 43, no. 3 (15 March 1957): 285–92.

Kohler, Robert E. *Lords of the Fly: Drosophila Genetics and the Experimental Life*. Chicago: University of Chicago Press, 1994.

Kopelman, Loretta. "Ethical Controversies in Medical Research: The Case of XYY Screening." *Perspectives in Biology and Medicine* 21 (Winter 1978): 196–204.

Kornberg, Roger. "Chromatin Structure: A Repeating Unit of Histones and DNA." *Science* 184 (1974): 868–71.

Kottler, Malcolm Jay. "From 48 to 46: Cytological Technique, Preconception, and the Counting of Human Chromosomes." *Bulletin of the History of Medicine* 48 (1974): 465–502.

Kowal, Emma. "Orphan DNA: Indigenous Samples, Ethical Biovalue and Postcolonial Science." *Social Studies of Science* 43 (2013): 577–97.

Kowal, Emma, and Joanna Radin. "Indigenous Biospecimen Collections and the Cryopolitics of Frozen Life." *Journal of Sociology* 51 (2015): 63–80.

Kraft, Alison. "Between Medicine and Industry: Medical Physics and the Rise of the Radioisotope, 1945–65." *Contemporary British History* 20 (2006): 1–35.

———. "Manhattan Transfer: Lethal Radiation, Bone Marrow Transplantation, and the Birth of Stem Cell Biology, ca. 1942–1961." *Historical Studies in the Natural Sciences* 39, no. 2 (2009): 171–218.

Krementsov, Nikolai. *Stalinist Science*. Princeton, NJ: Princeton University Press, 1997.

Krige, John. "Atoms for Peace, Scientific Internationalism, and Scientific Intelligence." In *Global Knowledge Power: Science and Technology in Interna-*

tional Affairs, vol. 21 of *Osiris*, edited by John Krige and Kai-Henrik Barth, 161–81. Chicago: University of Chicago Press, 2006.

Krimsky, S. *Genetic Alchemy: The Social History of the Recombinant DNA Controversy.* Cambridge, MA: MIT Press, 1982.

Kuhn, Thomas S. *The Structure of Scientific Revolutions.* 2nd ed. Chicago: University of Chicago Press, 1970.

Kutcher, Gerald. *Contested Medicine: Cancer Research and the Military.* Chicago: University of Chicago Press, 2009.

Lacadena, J. R. "Cytogenetics: Yesterday, Today and Forever—A Conceptual and Historic View." In *Chromosomes Today*, vol. 12, *Proceedings of the Sixth International Chromosome Conference Held in Helsinki, Finland, August 29–31, 1977*, edited by N. Henriques-Gil, J. S. Parker, and M. J. Puertas, 3–19. London: Chapman & Hall, 1997.

Lade, Quentin. "Une belle image pour une bonne revue: Une ethnographie des représentations visuelles en sciences expérimentales." *Genèses* 103, no. 2 (2016): 117–38.

Lambilliotte, M., ed. *Le mémorial officiel de l'Exposition Universelle et Internationale de Bruxelles.* Brussels: Établissements généraux d'Imprimerie, 1959.

Landecker, Hannah. *Culturing Life: How Cells Became Technologies.* Cambridge, MA: Harvard University Press, 2007.

———. "The Life of Movement: From Microcinematography to Live-Cell Imaging." *Journal of Visual Culture* 11, no. 3 (2012): 378–99.

———. "The Social as Signal in the Body of Chromatin." *Sociological Review Monographs* 64 (2016): 79–99.

Laplane, Robert. "Éloge de Raymond Turpin (1895–1988)." *Bulletin de l'Académie nationale de médecine* 173, no. 5 (1989): 535–43.

Largent, Mark A. *Breeding Contempt: The History of Coerced Sterilization in the United States.* New Brunswick, NJ: Rutgers University Press, 2011.

Latour, Bruno. *Science in Action: How to Follow Scientists and Engineers through Society.* Cambridge, MA: Harvard University Press, 1987.

Laurence, William L. "Drama of the Atomic Bomb Found Climax in July 16 Test." *New York Times*, 26 September 1945, 1 and 16.

Lederer, Susan E. *Subjected to Science: Human Experimentation in America before the Second World War.* Baltimore: Johns Hopkins University Press, 1995.

Ledley, Robert S. "Automatic Pattern Recognition for Clinical Medicine." *Proceedings of the IEEE* 57 (1969): 2017–35.

———. "High-Speed Automatic Analysis of Biomedical Pictures." *Science* 146, no. 3641 (1964): 216–23.

———. Preface to *Research Accomplishments, 1960–1970*, by National Biomedical Research Foundation, xv–xxi. Washington, DC: National Biomedical Research Foundation, 1973.

——. *Use of Computers in Biology and Medicine.* New York: McGraw-Hill, 1965.

Ledley, Robert S., and Frank H. Ruddle. "Automatic Analysis of Chromosome Karyograms." In *Mathematics and Computer Science in Biology and Medicine: Proceedings of Conference Held by the Medical Research Council in Association with the Health Departments, Oxford, July 1964,* edited by Medical Research Council, 189–209. London: Her Majesty's Stationery Office, 1965.

——. "Chromosome Analysis by Computer." *Scientific American* 214, no. 4 (1966): 40–46.

Leeming, William. "Tracing the Shifting Sands of 'Medical Genetics': What's in a Name?" *Studies in History and Philosophy of Biological and Biomedical Sciences* 41 (2010): 50–60.

Lejeune, Jérôme. "Le mongolisme, maladie chromosomique." *La Nature* 5296 (1959): 521–22.

——. "Scientific Impact of the Study of Fine Structure of Chromatids." In *Chromosome Identification: Techniques and Applications in Biology and Medicine; Proceedings of the 23rd Nobel Symposium Held September 25–27, 1972 at the Royal Academy of Sciences, Stockholm,* edited by T. Caspersson and L. Zech, 16–24. New York: Academic Press, 1973.

——. "The William Allan Memorial Award Lecture: On the Nature of Men." *American Journal of Human Genetics* 22, no. 2 (1970): 121–28.

Lejeune, Jérôme, Marthe Gautier, and Raymond Turpin. "Étude des chromosomes somatiques de neuf enfants mongoliens." *Comptes rendus hebdomadaires des séances de l'Académie des sciences* 248 (March 1959): 1721–22.

——. "Les chromosomes humains en culture de tissus." *Comptes rendus de l'Académie des sciences* 248 (26 January 1959): 602–3.

Lejeune, Jérôme, Raymond Turpin, and Marthe Gautier. "Le mongolisme, maladie chromosomique." *Bulletin de l'Académie nationale de médecine* 143 (14 April 1959): 256–65.

Lennox, Bernard. "Chromosomes for Beginners." *Lancet* 277, no. 7185 (1961): 1046–51.

Lenoir, Timothy, and Marguerite Hays. "The Manhattan Project for Biomedicine." In *Controlling Our Destinies,* edited by Peter Sloan, 29–62. Notre Dame, IN: University of Notre Dame Press, 2000.

Leonelli, Sabina. *Data-Centric Biology: A Philosophical Study.* Chicago: University of Chicago Press, 2016.

Leopold, Ellen. *Under the Radar: Cancer and the Cold War.* Brunswick, NJ: Rutgers University Press, 2008.

Levan, Albert. "Chromosome Studies on Some Human Tumors and Tissues of Normal Origin, Grown In Vivo and In Vitro at the Sloan-Kettering Institute." *Cancer* 9, no. 4 (1956): 648–63.

Lewontin, Richard C., Steven Rose, and Leon J. Kamin. *Not in Our Genes: Biology, Ideology, and Human Nature*. New York: Pantheon Books, 1984.

Lima-de-Faria, Antonio. *One Hundred Years of Chromosome Research and What Remains to Be Learned*. Dordrecht, The Netherlands: Kluwer Academic Publishers, 2003.

——. *Praise of Chromosome "Folly": Confessions of an Untamed Molecular Structure*. Singapore: World Scientific, 2008.

Lindee, M. Susan. "Genetic Disease in the 1960s: A Structural Revolution." *American Journal of Medical Genetics* 115 (2002): 75–82.

——. *Moments of Truth in Genetics and Medicine*. Baltimore: Johns Hopkins University Press, 2005.

——. *Suffering Made Real: American Science and the Survivors at Hiroshima*. Chicago: University of Chicago Press, 1994.

——. "Voices of the Dead: James Neel's Amerindian Studies." In *Lost Paradises and the Ethics of Research and Publication*, edited by Francisco M. Salzano and A. Magdalena Hurtado, 27–48. Oxford: Oxford University Press, 2004.

Lindee, M. Susan, and Joanna Radin. "Patrons of the Human Experience: A History of the Wenner-Gren Foundation for Anthropological Research, 1941–2016." *Current Anthropology* 57, supplement 14 (2016): S218–S301.

Lindee, M. Susan, and Ricardo Ventura Santos. "The Biological Anthropology of Living Human Populations: World Histories, National Styles, and International Networks: An Introduction to Supplement 5." *Current Anthropology* 53, supplement 5 (April 2012): S3–S16.

Lipphardt, Veronika. "'Geographical Distribution Patterns of Various Genes': Genetic Studies of Human Variation after 1945." *Studies in History and Philosophy of Biological and Biomedical Sciences* 47A (2014): 50–61.

——. "Isolates and Crosses in Human Population Genetics; or, A Contextualization of German Race Science." *Current Anthropology* 53, supplement 5 (2012): S69–S82.

Lloyd, D. C, and G. W. Dolphin. "Radiation-Induced Chromosome Damage in Human Lymphocytes." *British Journal of Industrial Medicine* 34 (1977): 261–73.

Lloyd, D. C., and R. J. Purrott. "Chromosome Aberration Analysis in Radiological Protection Dosimetry." *Radiation Protection Dosimetry* 1, no. 1 (1981): 19–28.

Lombardo, Paul A., ed. *A Century of Eugenics in America: From the Indiana Experiment to the Human Genome Project*. Bloomington: Indiana University Press, 2011.

——. "Tracking Chromosomes, Castrating Dwarves: Uninformed Consent and Eugenic Research." *Ethics & Medicine* 23 (2009): 149–64.

López-Beltrán, Carlos. "The Medical Origins of Heredity." In *Heredity Produced: At the Crossroads of Biology, Politics, and Culture, 1500–1870*, edited by Staffan Müller-Wille and Hans-Jörg Rheinberger, 105–32. Cambridge, MA: MIT Press, 2007.

Löwy, Ilana. "How Genetics Came to the Unborn: 1960–2000." *Studies in History and Philosophy of Biological and Biomedical Sciences* 47A (2014): 154–62.

———. *Imperfect Pregnancies: A History of Birth Defects and Prenatal Diagnosis.* Baltimore: Johns Hopkins University Press, 2017.

———. "Microscope Slides in the Life Sciences: Material, Epistemic and Symbolic Objects: Introduction." In "Microscope Slides: Reassessing a Neglected Historical Resource," edited by Ilana Löwy. Special issue of *History and Philosophy of the Life Sciences* 35, no. 3 (2013): 309–18.

Lunbeck, Elizabeth, and Lorraine Daston, eds. *Histories of Scientific Observation.* Chicago: University of Chicago Press, 2011.

Lynch, Michael, Simon A. Cole, Ruth McNally, and Kathleen Jordan. *Truth Machine: The Contentious History of DNA Fingerprinting.* Chicago: University of Chicago Press, 2008.

Lynch, Michael, and Steve Woolgar, eds. *Representation in Scientific Practice.* Cambridge, MA: MIT Press, 1990.

Lyon, Mary F. "Charles Edmund Ford, 24 October 1912–7 January 1999." *Biographical Memoirs of Fellows of the Royal Society* 47 (2001): 189–201.

Lyons, Richard D. "Genetic Abnormality Is Linked to Crime." *New York Times*, 21 April 1968, 1 and 72.

MacLean, Neil. "Sex Chromatin Surveys of Newborn Babies." In *The Sex Chromatin*, edited by Keith L. Moore, 202–10. Philadelphia: Saunders Company, 1966.

MacLean, N., W. M. Court Brown, Patricia A. Jacobs, D. J. Mantle, and J. A. Strong. "A Survey of Sex Chromatin Abnormalities in Mental Hospitals." *Journal of Medical Genetics* 5 (1968): 165–72.

MacLean, N., D. G. Harnden, and W. M. Court Brown. "Abnormalities of Sex Chromosome Constitution in Newborn Babies." *Lancet* 278, no. 7199 (19 August 1961): 406–8.

MacLean, N., J. M. Mitchell, D. G. Harnden, J. Williams, P. A. Jacobs, K. E. Buckton, A. G. Baikie, W. M. Court Brown, J. A. McBride, J. A. Strong, H. G. Close, and D. C. Jones. "A Survey of Sex-Chromosome Abnormalities among 4514 Mental Defectives." *Lancet* 279, no. 7224 (10 February 1962): 293–96.

Maienschein, Jane. *Embryos under the Microscope: The Diverging Meanings of Life.* Cambridge, MA: Harvard University Press, 2014.

Makino, Sajiro. *An Atlas of the Chromosome Numbers in Animals.* 2nd rev. ed. Ames: Iowa State College Press, 1951.

———. *Human Chromosomes.* Tokyo: Igaku Shoin, 1975.

Makino, Sajiro, and Motomochi Sasaki. "A Study of Somatic Chromosomes in a Japanese Population." *American Journal of Human Genetics* 13 (1961): 47–62.

Mark, Vernon H., and Frank R. Ervin. *Violence and the Brain.* New York: Harper & Row, 1970.

Marks, Lara V. *The Lock and Key of Medicine: Monoclonal Antibodies and the Transformation of Healthcare.* New Haven, CT: Yale University Press, 2015.

Martin, Aryn. "Can't Any Body Count? Counting as an Epistemic Theme in the History of Human Chromosomes." *Social Studies of Science* 34 (2004): 923–48.

Martin, Joseph D. *Solid State Insurrection: How the Science of Substance Made American Physics Matter.* Pittsburgh, PA: University of Pittsburgh Press, 2018.

Martínez-Patiño, María-José. "Personal Account: A Woman Tried and Tested." *Lancet* 366, supplement 1 (December 2005): S38.

Martynoga, Ben. *Molecular Tinkering: The Edinburgh Scientists Who Changed the Face of Modern Biology.* Kibworth Beauchamp, UK: Matador-Troubador Publishing, 2018.

Matlin, Karl S., Jane Maienschein, and Manfred D. Laublicher, eds. *Visions of Cell Biology: Reflections Inspired by Cowdry's "General Cytology."* Chicago: University of Chicago Press, 2018.

Mayr, Ernst. *Systematics and the Origin of Species.* New York: Columbia University Press, 1942.

Mazumdar, Pauline M. H. *Eugenics, Human Genetics and Human Failings: The Eugenics Society, Its Sources and Its Critics in Britain.* London: Routledge, 1992.

———, ed. *The Eugenics Movement: An International Perspective.* 6 vols. London: Routledge/Athena Press, 2007.

McGovern, Michael. "'The London/Baltimore Link Has Been Severed': The Economies of Human Gene Mapping and Mainframe Computing at the Moore Clinic, 1955–1973." MPhil diss., University of Cambridge, 2014.

M'Charek, Amade. *The Human Genome Diversity Project: An Ethnography of Scientific Practice.* Cambridge: Cambridge University Press, 2005.

McKusick, Victor A. "The Human Genome through the Eyes of a Clinical Geneticist." In "Human Gene Mapping 6: Oslo Conference (1981), Sixth International Workshop on Human Gene Mapping." Special issue of *Cytogenetics and Cell Genetics* 32 (1982): 7–23.

———. *Mendelian Inheritance in Man: Catalogs of Autosomal Dominant, Autosomal Recessive, and X-Linked Phenotypes.* Baltimore: Johns Hopkins University Press, 1983.

———. "A 60-Year Tale of Spots, Maps, and Genes." *Annual Review of Genomics and Human Genetics* 7 (2006): 1–27.

McKusick, Victor A., and Frank H. Ruddle. "Editorial: A New Discipline, a New Name, a New Journal." *Genomics* 1 (1987): 1–2.

McLeod, John, and Anne Johnstone. "The Secret Guinea Pigs: Parents Not Told of XYY Tests." *Evening Times (Glasgow)*, 13 June 1979, 1.

McRae, Donald. "Semenya Ready to Rip Up Record Books in Rio but Is the Track Level?" *Guardian*, 30 July 2016, 3–5.

Medical Research Council. *The Hazards to Man of Nuclear and Allied Radiations*. London: Her Majesty's Stationery Office, 1956. Cmnd. 9780.

———. *The Hazards to Man of Nuclear and Allied Radiations: A Second Report to the Medical Research Council*. London: Her Majesty's Stationery Office, 1960. Cmnd. 1225.

Mendelsohn, Mortimer L. Introduction to *Automation of Cytogenetics: Asilomar Workshop, Pacific Grove, California, November 30–December 2, 1976* [i.e., 1975], edited by Mortimer L. Mendelsohn, 1–2. Springfield, VA: National Technical Information Service, 1976.

Meselson, Matthew, and Franklin W. Stahl. "The Replication of DNA in *Escherichia coli*." *Proceedings of the National Academy of Sciences* 44 (1958): 671–82.

Migeon, Barbara R. *Females Are Mosaics: X Inactivation and Sex Differences in Disease*. Oxford: Oxford University Press, 2007.

Milam, Erika Lorraine. *Creatures of Cain: The Hunt for Human Nature in Cold War America*. Princeton, NJ: Princeton University Press, 2019.

———. "Men in Groups: Anthropology and Aggression, 1965–84." In *Scientific Masculinity*, vol. 30 of *Osiris*, edited by Erika Lorraine Milam and Robert A. Nye, 66–88. Chicago: University of Chicago Press, 2015.

Miller, Fiona. "'Your True and Proper Gender': The Barr Body as a *Good Enough* Science of Sex." *Studies in History and Philosophy of Biological and Biomedical Sciences* 37 (2006): 459–83.

Miller, James R. "Chromosome Registers—Problems and Perspectives." In *Population Cytogenetics: Studies in Humans; Proceedings of a Symposium on Human Population Cytogenetics Sponsored by the Birth Defects Institute of the New York State Department of Health, Held in Albany, New York, October 14–15, 1975*, edited by Ernest B. Hook and Ian H. Porter, 251–55. New York: Academic Press, 1977.

Miller, Larry. "What Becomes of the XYY Male?" *Lancet* 305, no. 7900 (25 January 1975): 221–22.

Miller, O. J., P. W. Allderdice, D. A. Miller, W. R. Breg, and B. R. Migeon. "Human Thymidine Kinase Gene Locus: Assignment to Chromosome 17 in a Hybrid of Man and Mouse Cells." *Science* 173 (1971): 244–45.

Mittwoch, Ursula. "The Chromosome Complement in a Mongolian Imbecile." *Annals of Eugenics* 17 (1952): 37–38.

———. *Sex Chromosomes*. New York: Academic Press, 1967.

Montagu, Ashley. "Chromosomes and Crime." *Psychology Today* 2, no. 5 (October 1968): 42–49.

———. *The Natural Superiority of Women.* New York: Macmillan, 1953.

Moore, Keith L. Introduction to *The Sex Chromatin,* edited by Keith L. Moore, 1–6. Philadelphia: Saunders Company, 1966.

———. "Sex Reversal in Newborn Babies." *Lancet* 273, no. 7066 (31 January 1959): 217–19.

———. "The Sexual Identity of Athletes." *Journal of the American Medical Association* 205 (1968): 787–88.

Moore, Keith L., and Murray L. Barr. "Smears from the Oral Mucosa in the Detection of Chromosomal Sex." *Lancet* 266, no. 6880 (9 July 1955): 57–58.

Moore, Kelly. *Disrupting Science: Social Movements, American Scientists, and the Politics of the Military, 1945–1975.* Princeton, NJ: Princeton University Press, 2008.

Morange, Michel. *A History of Molecular Biology.* Cambridge, MA: Harvard University Press, 1998.

Morgan, Lynn. *Icons of Life: A Cultural History of Human Embryos.* Berkeley: University of California Press, 2009.

Morgan, Thomas Hunt. "Has the White Man More Chromosomes Than the Negro?" *Science* 39 (14 June 1914): 827–28.

———. *Heredity and Sex.* New York: Columbia University Press, 1913.

Morton, Newton E. "Problems and Methods in the Genetics of Primitive Groups." *American Journal of Physical Anthropology* 28 (1968): 191–202.

Morton, Newton E., Patricia A. Jacobs, Anna Frackiewicz, Pamela Law, and Judith Hilditch. "The Effect of Structural Aberrations of the Chromosomes on Reproductive Fitness in Man." *Clinical Genetics* 8 (1975): 159–68.

Müller-Wille, Staffan, and Hans-Jörg Rheinberger. *A Cultural History of Heredity.* Chicago: University of Chicago Press, 2012.

Nasim, Omar W. *Observing by Hand: Sketching the Nebulae in the Nineteenth Century.* Chicago: University of Chicago Press, 2013.

National Academy of Sciences. *The Biological Effects of Atomic Radiation: A Report to the Public.* Washington, DC: National Academy of Sciences, National Research Council, 1956.

Navon, Daniel. *Mobilizing Mutations: Human Genetics in the Age of Patient Advocacy.* Chicago: University of Chicago Press, 2019.

Neel, James V. *Physician to the Gene Pool: Genetic Lessons and Other Stories.* New York: John Wiley & Sons, 1994.

———. "The Study of Natural Selection in Primitive and Civilized Human Populations." *Human Biology* 30 (1958): 43–72.

Neel, J. V., and F. M. Salzano. "A Prospectus for Genetic Studies on the American Indians." In *The Biology of Human Adaptability,* edited by Paul T. Baker and J. S. Weiner, 245–74. Oxford: Clarendon Press, 1966.

Neel, J. V., F. M. Salzano, P. C. Junqueira, F. Keiter, and D. Maybury-Lewis. "Studies on the Xavante Indians of the Brazilian Mato Grosso." *American Journal of Human Genetics* 16, no. 1 (1964): 52–140.

Neel, James V., and William J. Schull. *The Effect of Exposure to the Atomic Bombs on Pregnancy Termination in Hiroshima and Nagasaki.* Publication No. 461. Washington, DC: National Academy of Sciences, National Research Council, 1956.

Nelkin, Dorothy, and Laurence Tancredi. *Dangerous Diagnostics: The Social Power of Biological Information.* Chicago: University of Chicago Press, 1994.

"New Haven Conference (1973): First International Workshop on Human Gene Mapping." *Cytogenetics and Cell Genetics* 13 (1974): 1–216.

Nicolson, Malcolm, and John E. E. Fleming. *Imaging and Imagining the Fetus: The Development of Obstetric Ultrasound.* Baltimore: Johns Hopkins University Press, 2013.

November, Joseph. *Biomedical Computing: Digitizing Life in the United States.* Baltimore: John Hopkins University Press, 2012.

Nowell, P. C., and D. A. Hungerford. "Chromosome Studies on Normal and Leukemic Human Leukocytes." *Journal of the National Cancer Institute* 25 (1960): 85–93.

Ohno, Susumu. "Single-X Derivation of Sex Chromatin." In *The Sex Chromatin*, edited by Keith L. Moore, 113–28. Philadelphia: W. B. Saunders Company, 1966.

Ohno, Susumu, and Sajiro Makino. "The Single-X Nature of Sex Chromatin in Man." *Lancet* 277, no. 7168 (14 January 1961): 78–79.

Olby, Robert. *Francis Crick: Hunter of Life's Secrets.* Cold Spring Harbor, NY: Cold Spring Harbor Laboratory Press, 2009.

Olins, Donald E., and Ada L. Olins. "Chromatin History: Our View from the Bridge." *Nature Reviews/Molecular Cell Biology* 4 (October 2003): 809–14.

O'Malley, Maureen, and Staffan Müller-Wille, eds. "The Cell as Nexus: Connections between the History, Philosophy and Science of Cell Biology." *Studies in History and Philosophy of Biological and Biomedical Sciences* 41, no. 3 (2010): 169–306.

Painter, Theophilus S. "A Comparative Study of the Chromosomes of Mammals." *American Naturalist* 59 (1925): 385–409.

———. "The Sex Chromosomes of Man." *American Naturalist* 58, no. 659 (1924): 506–24.

———. "Studies in Mammalian Spermatogenesis II: The Spermatogenesis of Man." *Journal of Experimental Zoology* 37 (1923): 291–335.

Panofsky, Aaron. *Misbehaving Science: Controversy and Development of Behavior Genetics.* Chicago: University of Chicago Press, 2014.

Pappworth, Maurice H. *Human Guinea Pigs.* London: Routledge & Kegan Paul, 1967.

Pardue, Mary-Lou, and Joseph G. Gall. "Molecular Hybridization of Radioactive DNA to the DNA of Cytological Preparations." *Proceedings of the National Academy of Sciences* 64 (1969): 600–604.

Park, W. Wallace. "The YY Syndrome" (letter to the editor). *Lancet* 288, no. 7479 (31 December 1966): 1468.

Patau, Klaus. "Chromosome Identification and the Denver Report." *Lancet* 227, no. 7183 (1961): 933–34.

———. "The Identification of Individual Chromosomes, Especially in Man." *American Journal of Human Genetics* 12 (1960): 250–76.

Paul, Diane B. *Controlling Human Heredity, 1865 to the Present*. Amherst, NY: Humanity Books, 1998.

———. "'Our Load of Mutations' Revisited." *Journal of the History of Biology* 20 (1987): 321–35.

Paul, Diane B., and Jeffrey P. Brosco. *The PKU Paradox: A Short History of a Genetic Disease*. Baltimore: Johns Hopkins University Press, 2013.

Paul, Diane B., and Paul J. Edelson. "The Struggle over Metabolic Screening." In *Molecularizing Biology and Medicine: New Practices and Alliances, 1910s–1970s*, edited by Soraya de Chadarevian and Harmke Kamminga, 203–20. Amsterdam: Harwood Academic Publishers, 1998.

Pauling, Linus, Harvey A. Itano, S. J. Singer, and Ibert C. Wells. "Sickle Cell Anemia, a Molecular Disease." *Science* 110 (1949): 543–48.

Penrose, Lionel S. *A Clinical and Genetic Study of 1280 Cases of Mental Defect*. London: His Majesty's Stationery Office, 1938.

———. "Human Chromosomes, Normal and Aberrant." *Proceedings of the Royal Society of London B* 164 (1966): 311–19.

———. Introduction to "New Aspects of Human Genetics," edited by Charles E. Ford and Harry Harris. Special issue of *British Medical Bulletin* 25 (1969): 1–4.

———. "Introductory Address." In *Chicago Conference: Standardization in Human Cytogenetics. Sponsored by the National Foundation–March of Dimes at the University of Chicago Center for Continuing Education, September 3, 4 and 10, 1966*, edited by Daniel Bergsma, John L. Hamerton, and Harold P. Klinger. Birth Defects: Original Article Series, vol. 2, no. 2, 1–2. New York: Liss, 1966.

———. "The London Conference on 'The Normal Human Karyotype', 28–30 August 1963." *Annals of Human Genetics* 27 (1964): 295–98.

———. "Maternal Age, Order of Birth and Developmental Abnormalities." *Journal of Mental Science* 85 (1939): 1141–50.

———. "A Note on the Measurements of Human Chromosomes." *Annals of Human Genetics* 28 (1964): 195–96.

———. "A Proposed Standard System of Nomenclature of Human Mitotic Chromosomes (Denver, Colorado). Editorial Comment." *Annals of Human Genetics* 24 (1960): 319.

Pickstone, John V. *Ways of Knowing: A New History of Science, Technology and Medicine.* Chicago: University of Chicago Press, 2001.

Piller, Gordon. *Rays of Hope—The Story of the Leukaemia Research Fund.* London: Leukaemia Research Fund, 1994.

Piper, J. "Cytoscan." *MRC News,* no. 56 (September 1992): 26–27.

Piper, J., E. Granum, D. Rutovitz, and H. Ruttledge. "Automation of Chromosome Analysis." *Signal Processing* 2 (1980): 203–21.

Pitt, Joseph C. "The Epistemology of the Very Small." In *Discovering the Nanoscale,* edited by Davis Baird, Alfred Nordmann, and Joachim Schummer, 157–63. Amsterdam: IOS Press, 2004.

Polani, Paul E. "Human and Clinical Cytogenetics: Origins, Evolution and Impact." *European Journal of Human Genetics* 5 (1997): 117–28.

——. "The Incidence of Chromosomal Malformations." *Proceedings of the Royal Society of Medicine* 63 (1970): 50–52.

——. "John Alexander Fraser Roberts, 8 September 1899–15 January 1987." *Biographical Memoirs of Fellows of the Royal Society* 38 (1992): 307–21.

Polani, Paul E., Eva D. Alberman, Benjamin Alexander, Philip F. Benson, A. Caroline Berry, Sarah C. Blunt, Michael G. Daker, Anthony H. Fensom, Donald M. Garrett, Valerie M. McGuire, John Alexander Fraser Roberts, Mary J. Seller, and Jack D. Singer. "Sixteen Years' Experience of Counselling, Diagnosis, and Prenatal Detection in One Genetic Centre: Progress, Results, and Problems." *Journal of Medical Genetics* 16 (1979): 166–75.

Polani, P. E., J. H. Briggs, C. E. Ford, C. M. Clarke, and J. M. Berg. "A Mongol Girl with 46 Chromosomes." *Lancet* 275, no. 7127 (2 April 1960): 721–24.

Polani, Paul E., M. H. Lessop, and P. M. F. Bishop. "Colour-Blindness in 'Ovarian Agenesis' (Gonadal Dysplasia)." *Lancet* 268, no. 6934 (21 July 1956): 118–20.

Poole, Robert. *Earthrise: How Man First Saw the Earth.* New Haven, CT: Yale University Press, 2008.

Porter, Theodore. *Genetics in the Madhouse: The Unknown History of Human Heredity.* Princeton, NJ: Princeton University Press, 2018.

——. *Trust in Numbers: The Pursuit of Objectivity in Science and Public Life.* Princeton, NJ: Princeton University Press, 1996.

Price, W. H., J. A. Strong, P. B. Whatmore, and W. F. McClemont. "Criminal Patients with XYY Sex-Chromosome Complement." *Lancet* 287, no. 7437 (12 March 1966): 565–66.

Price, W. H., and P. B. Whatmore. "Behaviour Disorders and Pattern of Crime among XYY Males Identified at a Maximum Security Hospital." *British Medical Journal* 1 (4 March 1967): 533–36.

——. "Criminal Behaviour and the XYY Male." *Nature* 213 (25 February 1967): 815.

Probert, M., and C. Rawlings. "Overview of GGM11." In "Human Gene Mapping 11: London Conference (1991), Eleventh International Workshop on Human Gene Mapping." Special issue of *Cytogenetics and Cell Genetics* 58 (1991): 2.

Puck, Theodore T. "Living History Biography." *American Journal of Medical Genetics* 53 (1994): 274–84.

Pyeritz, Reed, Jon Beckwith, and Larry Miller. "XYY Disclosure Condemned (Letter)." *New England Journal of Medicine* 293 (4 September 1975): 508.

Rader, Karen. *Making Mice: Standardizing Animals for American Biomedical Research, 1900–1955*. Princeton, NJ: Princeton University Press, 2004.

Radin, Joanna. "Latent Life: Concepts and Practices of Tissue Preservation in the International Biological Program." *Social Studies of Science* 43 (2013): 483–508.

———. *Life on Ice: A History of New Uses for Cold Blood*. Chicago: University of Chicago Press, 2017.

———. "Unfolding Epidemiological Stories: How the WHO Made Frozen Blood a Flexible Resource for the Future." *Studies in History and Philosophy of Biological and Biomedical Sciences* 47A (2014): 62–73.

Radin, Joanna, and Emma Kowal. "Indigenous Blood and Ethical Regimes in the United States and Australia since the 1960s." *American Ethnologist* 42, no. 4 (2015): 749–65.

Ramsden, Edmund. "Surveying the Meritocracy: The Problems of Intelligence and Mobility in the Studies of the Population Investigation Committee." *Studies in History and Philosophy of Biological and Biomedical Sciences* 47A (2014): 130–41.

Rapp, Rayna. *Testing Women, Testing the Fetus: The Social Impact of Amniocentesis in America*. New York: Routledge, 1999.

Rasmussen, Nicolas. *Picture Control: The Electron Microscope and the Transformation of Biology in America, 1940–1960*. Stanford, CA: Stanford University Press, 1997.

Ratcliffe, Shirley G., and D. G. Axworthy. "What Is to Be Done with the XYY Fetus?" (letter). *British Medical Journal* 2, no. 6191 (15 September 1979): 672.

Ratcliffe, Shirley G., Lynne Murray, and Peter Teague. "Edinburgh Study of Growth and Development of Children with Sex Chromosome Abnormalities III." *Birth Defects: Original Article Series* 22, no. 3 (1986): 73–118.

Ratcliffe, Shirley G., and Natalie Paul, eds. *Prospective Studies on Children with Sex Chromosome Aneuploidy: Reports from a Workshop Sponsored by the March of Dimes Birth Defects Foundation, Held in Edinburgh, June 1984*. Birth Defects: Original Article Series, vol. 22, no. 3. New York: Liss, 1986.

Rawlings, C. J., C. Brunn, S. Bryant, R. J. Robbins, and R. E. Lucier. "Report of the Informatics Committee." *Cytogenetics and Cell Genetics* 58 (1991): 1833–38.

Reardon, Jenny. *The Postgenomic Condition: Ethics, Justice, and Knowledge.* Chicago: University of Chicago Press, 2017.

———. *Race to the Finish: Identity and Governance in the Age of Genomics.* Princeton, NJ: Princeton University Press, 2004.

Redding, Audrey, and Kurt Hirschhorn. *Guide to Human Chromosome Defects.* Birth Defects: Original Article Series, vol. 4, no. 4. New York: National Foundation, March of Dimes, 1968.

Rheinberger, Hans-Jörg. *An Epistemology of the Concrete: Twentieth-Century Histories of Life.* Durham, NC: Duke University Press, 2010.

———. "Putting Isotopes to Work: Liquid Scintillation Counters, 1950–1970." In *Instrumentation between Science, State and Industry*, edited by Bernward Joerges and Terry Shinn, 143–74. Dordrecht, The Netherlands: Kluwer, 2001.

———. *Toward a History of Epistemic Things: Synthesizing Proteins in the Test Tube.* Stanford, CA: Stanford University Press, 1997.

Rheinberger, Hans-Jörg, and Jean-Paul Gaudillière, eds. *Classical Genetic Research and Its Legacy: The Mapping Cultures of Twentieth-Century Genetics.* London: Routledge, 2004.

Rheinberger, Hans-Jörg, and Staffan Müller-Wille, eds. *The Gene: From Genetics to Postgenomics.* Chicago: University of Chicago Press, 2017.

Richardson, Sarah S. *Sex Itself: The Search for Male and Female in the Human Genome.* Chicago: University of Chicago Press, 2013.

Richardson, Sarah S., and Hallam Stevens, eds. *Postgenomics: Perspectives on Genomics after the Genome.* Durham, NC: Duke University Press, 2015.

Richmond, Marsha L. "Women as Public Scientists in the Atomic Age: Rachel Carson, Charlotte Auerbach, and Genetics." *Historical Studies in the Natural Sciences* 47, no. 3 (2017): 349–88.

Robertson Smith, David, and William M. Davidson, eds. *Symposium on Nuclear Sex.* London: William Heinemann Medical Books, 1958.

Robinson, Arthur. "Living History: An Autobiography of Arthur Robinson." *American Journal of Medical Genetics* 35 (1990): 475–80.

Robinson, Arthur, Herbert A. Lubs, and Daniel Bergsma, eds. *Sex Chromosome Aneuploidy: Prospective Studies on Children.* Birth Defects: Original Article Series, vol. 15, no. 1. New York: Liss, 1979.

Roblin, Richard. "The Boston XYY Case: Controversial Experiment Ended Controversially." *Hastings Center Report* 5, no. 4 (1975): 5–8.

Rodríguez-Hernández, M. Louisa, and Eladio Montoya. "Fifty Years of Evolution of the Term Down's Syndrome." *Lancet* 378, no. 9789 (July 2011): 402.

Roff, Sue Rabbitt. *Hotspots: The Legacy of Hiroshima and Nagasaki.* London: Cassell, 1995.

Roll-Hansen, Nils. *The Lysenko Effect: The Politics of Science.* Amherst, NY: Humanity, 2005.

Rostand, Jean. "Parviendra-t-on bientot a guérir les enfants atteints de mongolisme?" *Le Figaro litteraire du samedi*, 5 September 1959.

Royce, Kenneth. *The XYY Man*. London: Hodder & Stoughton, 1970.

Ruddle, Frank H. "Quantitation and Automation of Chromosomal Data with Special Reference to the Chromosomes of the Ham[p]shire Pig (*Sus scrofa*)." In *Cytogenetics of Cells in Culture*, vol. 3 of *Symposium of the International Society for Cell Biology*, edited by R. J. C. Harris, 273–305. New York: Academic Press, 1964.

Ruddle, F. H., V. M. Chapman, F. Ricciuti, M. Murnane, R. Klebe, and P. Meera Khan. "Linkage Relationships of 17 Human Gene Loci as Determined by Man-Mouse Somatic Cell Hybrids." *Nature–New Biology* 232, no. 29 (1971): 69–73.

Ruddle, Frank H., and Kenneth K. Kidd. "First Human Gene Mapping Interim Meeting—New Haven, 1988." In "Human Gene Mapping 9.5: New Haven Conference (1988), update to the Ninth International Workshop on Human Gene Mapping." Special issue of *Cytogenetics and Cell Genetics* 49 (1988): 1.

———. "The Human Gene Mapping Workshops in Transition." In "Human Gene Mapping 10: New Haven Conference (1989), Tenth International Workshop on Human Gene Mapping." Special issue of *Cytogenetics and Cell Genetics* 51 (1989): 1–2.

Rutovitz, Denis. "Reflections on the Past, Present and Future of Automated Aberration Scoring Systems for Radiation Damage." *Journal of Radiation Research* 33, supplement (1992): 1–30.

Rutovitz, Denis, D. K. Green, A. S. J. Farrow, and D. C. Mason. "Computer-Assisted Measurement in the Cytogenetic Laboratory." In *Pattern Recognition*, edited by Bruce G. Batchelor, 303–29. New York: Plenum Publishing, 1978.

Sachs, L., M. Danon, M. Feldman, and D. M. Serr. "The Prenatal Diagnosis of Human Abnormalities." *Acta Genetica et Statistica Medica* 6 (1956): 254–55.

Sachse, Carola. "Ein 'als Neugründung zu deutender Beschluss': vom Kaiser-Wilhelm Institut für Anthropologie, menschliche Erblehre und Eugenik zum Max-Planck-Institut für molekulare Genetik." *Medizinhistorisches Journal* 46, no. 2011 (2011): 24–50.

Saitoh, Yasushi, and Ulrich K. Laemmli. "Metaphase Chromosome Structure: Bands Arise from a Differential Folding Path of the Highly AT-Rich Scaffold." *Cell* 76 (1994): 609–22.

Sandberg, Avery A., George F. Koepf, Takaaki Ishihara, and Theodore S. Hauschka. "An XYY Human Male." *Lancet* 278, no. 7200 (26 August 1961): 488–89.

Santesmases, María Jesús. "The Biological Landscape of Polyploidy: Chromosomes under Glass in the 1950s." *History and Philosophy of the Life Sciences* 35 (2013): 91–97.

——. "Cereals, Chromosomes and Colchicine: Crop Varieties at the Estación Experimental Aula Dei and Human Cytogenetics, 1948–1958." In *Human Heredity in the Twentieth Century*, edited by Bernd Gausemeier, Staffan Müller-Wille, and Edmund Ramsden, 127–40. London: Pickering and Chatto, 2013.

——. "Circulating Biomedical Images: Bodies and Chromosomes in the Post-Eugenic Era." *History of Science* 55 (2017): 395–430.

——. "The Human Autonomous Karyotype and the Origins of Prenatal Testing: Children, Pregnant Women and Early Down's Syndrome Cytogenetics, Madrid, 1962–1975." *Studies in History and Philosophy of Biological and Biomedical Sciences* 47A (2014): 142–53.

——. "Human Chromosomes and Cancer: Tumors and the Geographies of Cytological Practices, 1951–1956." *Historical Studies in the Natural Sciences* 45 (2015): 85–114.

——. "Size and the Centromere: Translocations and Visual Cultures in Early Human Genetics." In *Making Mutations: Objects, Practices, Contexts* (preprint no. 393), edited by Luis Campos and Alexander von Schwerin, 189–207. Berlin: Max Planck Institute for the History of Science, 2010.

Santesmases, María Jesús, and Edna Suárez-Díaz, eds. "A Cell-Based Epistemology: Human Genetics in the Era of Biomedicine." Special issue of *Historical Studies in the Natural Sciences* 45, no. 1 (2015): 1–197.

Santos, Ricardo Ventura. "Indigenous People, Postcolonial Contexts and Genomic Research in the Late Twentieth Century: A View from Amazonia (1960–2000)." *Critique of Anthropology* 22 (2002): 81–104.

——. "Indigenous Peoples, Bioanthropological Research, and Ethics in Brazil: Issues in Participation and Consent." In *The Nature of Difference: Science, Society and Human Biology*, edited by George T. H. Ellison and Alan H. Goodman, 181–202. Boca Raton, FL: Taylor & Francis, 2006.

Santos, Ricardo Ventura, Susan Lindee, and Vanderlei Sebastião de Souza. "Varieties of the Primitive: Human Biological Diversity Studies in Cold War Brazil (1962–1970)." *American Anthropologist* 116 (2014): 723–35.

Sapp, Jan. *Beyond the Gene: Cytoplasmic Inheritance and the Struggle for Authority in Genetics*. Oxford: Oxford University Press, 1987.

Satzinger, Helga. *Differenz und Vererbung: Geschlechterordnungen in der Genetik und der Hormonforschung 1890–1950*. Vienna: Böhlau Verlag, 2009.

Scheffler, Robin Wolfe. *A Contagious Cause: The American Hunt for Cancer Viruses and the Rise of Molecular Medicine*. Chicago: University of Chicago Press, 2019.

——. "Managing the Future: The Special Virus Leukemia Program and Acceleration of Biomedical Research." *Studies in History and Philosophy of Biological and Biomedical Sciences* 48 (2014): 231–49.

Schickore, Jutta. *The Microscope and the Eye: A History of Reflections, 1740–1870.* Chicago: University of Chicago Press, 2007.

Schmuhl, Hans-Walter. *The Kaiser Wilhelm Institute for Anthropology, Human Heredity and Eugenics, 1927–1945: Crossing Boundaries.* Dordrecht, The Netherlands: Springer, 2008.

Schneider, William H. "Blood Group Research in Great Britain, France and the United States between the World Wars." *Yearbook of Physical Anthropology* 38 (1995): 87–114.

———. "The History of Research on Blood Group Genetics: Initial Discovery and Diffusion." *History and Philosophy of the Life Sciences* 18 (1996): 277–303.

Schroeder-Gudehus, B., and D. Cloutier. "Popularizing Science and Technology during the Cold War: Brussels 1958." In *Fair Representations: World's Fairs and the Modern World*, edited by R. W. Rydell and N. Gwin, 157–80. Amsterdam: VU Press, 1994.

Semendeferi, Ioanna. "Legitimating a Nuclear Critic: John Gofman, Radiation Safety, and Cancer Risks." *Historical Studies in the Natural Sciences* 38 (2008): 259–301.

Sever, Lowell E. "A Conversation with Jack Schull." *Epidemiology* 15, no. 1 (2004): 118–22.

"Sex Tests in Search of Olympic Supermen . . ." *Daily Express*, 4 October 1968.

Simpson, Joe Leigh, Arne Ljungqvist, Malcolm A. Ferguson-Smith, Albert de la Chapelle, Louis J. Elsas, A. A. Ehrhardt, Myron Genel, Elizabeth A. Ferris, and Alison Carlson. "Gender Verification in the Olympics." *Journal of the American Medical Association* 284, no. 12 (2000): 1568–69.

Skolnick, M. H., and U. Francke. "Report of the Committee on Human Gene Mapping by Recombinant DNA Techniques." Human Gene Mapping 6: Oslo Conference (1981), Sixth International Workshop on Human Gene Mapping. *Cytogenetics and Cell Genetics* 32 (1982): 194–204.

Smith, Peter G. "The 1957 MRC Report on Leukaemia and Aplastic Anaemia in Patients Irradiated for Ankylosing Spondylitis." *Journal of Radiological Protection* 27, special supplement (2007): B3–B14.

Smocovitis, Vassiliki Betty. "Genetics behind Barbed Wire: Masuo Kodani, Émigré Geneticists, and Wartime Genetics Research at Manzanar Relocation Center." *Genetics* 187, no. 2 (2011): 357–66.

Solomon, E., and W. F. Bodmer. Introduction to "Human Gene Mapping 11: London Conference (1991), Eleventh International Workshop on Human Gene Mapping." Special issue of *Cytogenetics and Cell Genetics* 58 (1991): 1.

Sontag, Susan. *On Photography.* New York: Farrar, Straus & Giroux, 1977.

Soudek, D. "Chromosomal Variants with Normal Phenotype in Man." *Journal of Human Evolution* 2 (1973): 341–55.

Southern, E. M. "Application of DNA Analysis to Mapping the Human Genome." In "Human Gene Mapping 6: Oslo Conference (1981), Sixth International Workshop on Human Gene Mapping." Special issue of *Cytogenetics and Cell Genetics* 32 (1982): 52–57.

Sperling, Karl. "50 Jahre Max-Planck-Institut für molekulare Genetik—Die Wende zur Humangenetik." In *Gene und Menschen: 50 Jahre Forschung am Max-Planck-Institut für molekulare Genetik*, edited by Martin Vingron, 76–87. Berlin: Max-Planck Institut für molekulare Genetik, 2014.

"Spotting Flaws in Genes at Birth." *Medical World News*, 22 November 1968, 23.

Stannard, J. Newell. *Radioactivity and Health: A History*. Springfield, VA: Office of Scientific and Technological Information, 1988.

Steele, Mark W., and W. Roy Breg. "Chromosome Analysis of Human Amniotic Fluid Cells." *Lancet* 287, no. 7434 (1966): 383–85.

Stern, Alexandra Minna. *Eugenic Nation: Faults and Frontiers of Better Breeding in Modern America*. Oakland: University of California Press, 2005.

———. *Telling Genes: The Story of Genetic Counseling in America*. Baltimore: Johns Hopkins University Press, 2012.

Stevens, Hallam. *Life out of Sequence: A Data Driven History of Bioinformatics*. Chicago: University of Chicago Press, 2013.

Stevenson, Alan C., Harold A Johnston, M. I. Patricia Stewart, and Douglas R. Golding. "Congenital Malformations: A Report of a Study of a Series of Consecutive Births in 24 Centres." *Bulletin of the World Health Organization* 34, supplement (1966).

Stewart, Donald A., ed. *Children with Sex Chromosome Aneuploidy: Follow-up Studies. Proceedings of a Conference Held in Toronto, Ontario, Canada, Sponsored by the March of Dimes Birth Defects Foundation*. Birth Defects: Original Article Series, vol. 18, no. 4. New York: Liss, 1982.

Strasser, Bruno J. *Collecting Experiments: Making Big Data Biology*. Chicago: University of Chicago Press, 2019.

———. "The Experimenter's Museum: GenBank, Natural History, and the Moral Economy of Biomedicine." *Isis* 102 (2011): 60–96.

Strasser, Bruno J., and Soraya de Chadarevian. "The Comparative and the Exemplary: Revisiting the Early History of Molecular Biology." *History of Science* 49 (2011): 317–36.

Strickland, Stephen P. *Politics, Science, and Dread Disease: A Short History of United States Medical Research Policy*. Cambridge, MA: Harvard University Press, 1972.

Sturdy, Steve, and Roger Cooter. "Science, Scientific Management, and the Transformation of Medicine in Britain c. 1870–1950." *History of Science* 36 (1998): 421–66.

Suárez-Díaz, Edna, Vivette García-Deister, and Emily E. Vasquez. "Populations of Cognition: Practices of Inquiry into Human Populations in Latin America." *Perspectives on Science* 25 (2017): 551–63.

Sumner, A. T., and A. C. Chandley, eds. *Chromosomes Today*, vol. 1, *Proceedings of the 11th International Chromosome Conference Held in Edinburgh, UK, August 1992*. London: Chapman & Hall, 1993.

Taussig, Karen-Sue. *Ordinary Genomes: Science, Citizenship, and Genetic Identities*. Durham, NC: Duke University Press, 2009.

Taylor, J. Herbert. "Sister Chromatid Exchanges in Tritium-Labeled Chromosomes." *Genetics* 43 (1958): 515–29.

Taylor, J. Herbert, Philip S. Woods, and Walter L. Hughes. "The Organization and Duplication of Chromosomes as Revealed by Autoradiographic Studies Using Tritium-Labeled Thymidine." *Proceedings of the National Academy of Sciences* 43 (1957): 122–28.

te Heesen, Anke. *The Newspaper Clipping: A Modern Paper Object*. Manchester, UK: Manchester University Press, 2014.

Thomson, Arthur Landsborough. *Origins and Policy of the Medical Research Council*. Vol. 1 of *Half a Century of Medical Research*. London: Her Majesty's Stationery Office, 1973.

——. *The Programme of the Medical Research Council*. Vol. 2 of *Half a Century of Medical Research*. London: Her Majesty's Stationery Office, 1975.

Timmermans, Stefan, and Mara Buchbinder. *Saving Babies? The Consequences of Newborn Genetic Screening*. Chicago: University of Chicago Press, 2012.

Timmermans, Stefan, and Steven Epstein. "A World of Standards but Not a Standard World: Toward a Sociology of Standards and Standardization." *Annual Review of Sociology* 36 (2010): 69–89.

Tjio, Joe Hin, and Albert Levan. "The Chromosome Number of Man." *Hereditas* 42 (26 January 1956): 1–6.

Tjio, Joe Hin, and Theodore T. Puck. "The Somatic Chromosomes of Man." *Proceedings of the National Academy of Sciences* 44 (1958): 1229–37.

Tough, Ishbel M., and W. Michael Court Brown. "Chromosome Aberrations and Exposure to Ambient Benzene." *Lancet* 285, no. 7387 (27 March 1965): 684.

"Tracing Genetic Diseases." *Computing Report for the Scientist and the Engineer* 3, no. 6 (November 1967): 7–9.

Trautner, Thomas A. "'Ich hätte mir gar nichts anderes vorstellen können.'" In *Gene und Menschen: 50 Jahre Forschung am Max-Planck-Institut für molekulare Genetik*, edited by Martin Vingron, 62–71. Berlin: Max-Planck Institut für molekulare Genetik, 2014.

Turda, Marius, and Aaron Gillette. *Latin Eugenics in Comparative Perspective*. London: Bloomsbury, 2014.

Turpin, Raymond, and Alexandre Caratzali. "Remarques sur les ascendants et les collatéraux des sujets atteints du mongolisme." *Presse Médicale* 59 (25 July 1934): 1186–90.

Turpin, R., A. Caratzali, and H. Rogier. "Étude étiologique de 104 cas de mongolisme et considérations sur la pathogénie de cette maladie." In *Congrès latin d'eugénique: Report*, edited by Fédération internationale latin des sociétés d'eugénique, 154–70. Paris: Masson, 1937.

Turrini, Mauro. "Continuous Grey Scales versus Sharp Contrasts: Styles of Presentation in Italian Clinical Cytogenetics." *Human Studies* 35 (2012): 1–25.

———. "The Controversial Molecular Turn in Prenatal Diagnosis: CGH-Array Clinical Approaches and Biomedical Platforms." *Tecnoscienza—Italian Journal of Science & Technology Studies* 5 (2014): 115–39.

Urofsky, Melvin I. "'Among the Most Humane Moments in All Our History': *Brown v. Board of Education* in Historical Perspective." In *Black, White, and Brown: The Landmark School Desegregation Case in Retrospect*, edited by Clare Cushman and Melvin I. Urofsky, 1–45. Washington, DC: CQ Press, 2004.

Veatch, Robert M. "The Unexpected Chromosome . . . : A Counselor's Dilemma." *Hastings Center Report* 2, no. 1 (February 1972): 8–9.

Vogel, Friedrich, and Arno G. Motulsky. *Human Genetics: Problems and Approaches*. 2nd rev. ed. Berlin: Springer Verlag, 1986.

Walker, J. Samuel. *Permissible Dose: A History of Radiation Protection in the Twentieth Century*. Berkeley: University of California Press, 2000.

Washburn, S. L. "The New Physical Anthropology." *Transactions of the New York Academy of Sciences* 13 (1951): 298–304.

Washington, Harriet A. "Born for Evil? Stereotyping the Karyotype: A Case History in the Genetics of Aggressiveness." In *Twentieth Century Ethics of Human Subjects Research: Historical Perspectives on Values, Practices, and Regulations*, edited by Volker Roelke and Giovanni Maio, 319–33. Stuttgart: Franz Steiner Verlag, 2004.

Watson, James D. Foreword to *Chromosome Structure and Function*, edited by James D. Watson, xv. Cold Spring Harbor Symposia on Quantitative Biology 38. Cold Spring Harbor, NY: Cold Spring Harbor Laboratory, 1974.

———. *Molecular Biology of the Gene*. New York: W. A. Benjamin, 1965.

Watson, J. D., and F. H. C. Crick. "Genetical Implications of the Structure of Deoxyribonucleic Acid." *Nature* 171 (1953): 964–67.

Weidman, Nadine. "Popularizing the Ancestry of Man: Robert Ardrey and the Killer Instinct." *Isis* 102 (2011): 269–99.

Weiner, J. S. *A Guide to the Human Adaptability Proposals, with a Contribution by Paul T. Baker*. 2nd ed. IBP Handbook no. 1. Oxford: Blackwell Scientific Publications, 1969.

Weiner, J. S., and J. A. Lourie. *Human Biology: A Guide to Field Methods*. IBP Handbook no. 9. Oxford: Blackwell Scientific Publications, 1969.

Weiner, Jonathan. *Time, Love, Memory: A Great Biologist and His Quest for the Origins of Behavior.* London: Faber, 1999.

Wexler, Alice. *Mapping Fate: A Memoir of Family, Risk, and Genetic Research.* New York: Random House, 1995.

"What Is to Be Done with the XYY Fetus?" (editorial). *British Medical Journal* 1, no. 6177 (9 June 1979): 1519–20.

"When Atom Bomb Struck—Uncensored." *Life* 33, no. 13 (29 September 1952): 19–25.

Wilson, Duncan. *The Making of British Bioethics.* Manchester, UK: Manchester University Press, 2014.

Wise, Norton. *The Values of Precision.* Princeton, NJ: Princeton University Press, 1995.

World Health Organization. "Genetics and Your Health." *World Health: The Magazine of the World Health Organization,* September 1966.

———. "Indigenous People and Participatory Health Research." http://www.who .int/ethics/indigenous_peoples/en/index1.html.

———. *Research in Population Genetics of Primitive Groups.* Technical Report No. 279. Geneva: World Health Organization, 1964. https://apps.who.int/iris/ bitstream/handle/10665/40586/9241202793_eng.pdf.

———. *Research on Human Population Genetics: Report of a WHO Scientific Group.* Technical Report No. 387. Geneva: World Health Organization, 1968. http://whqlibdoc.who.int/trs/WHO_TRS_387.pdf.

———. *The Second Ten Years of the World Health Organization, 1958–1967.* Geneva: World Health Organization, 1968.

World Health Organization, Expert Committee on Health Statistics. *Epidemiological Methods in the Study of Chronic Diseases.* Technical Report No. 365. Geneva: World Health Organization, 1967.

Worthington, E. B., ed. *The Evolution of IBP.* Cambridge: Cambridge University Press, 1975.

Wright, S. *Molecular Politics: Developing American and British Regulatory Policy for Genetic Engineering, 1972–1982.* Chicago: University of Chicago Press, 1994.

Yoxen, E. "Giving Life a New Meaning: The Rise of the Molecular Biology Establishment." In *Scientific Establishments and Hierarchies,* edited by N. Elias, H. Martins, and R. Whitley, 123–43. Sociology of the Sciences Yearbook 6. Dordrecht, The Netherlands: Reidel, 1982.

Yu, J., J. Xiao, X. Ren, K. Lao, and X. S. Xie. "Probing Gene Expression in Live Cells, One Protein at a Time." *Science* 311, no. 5767 (17 March 2006): 1600–1603.

"The YY Syndrome" (editorial). *Lancet* 287, no. 7437 (12 March 1966): 583–84.

Zallen, D. T. "Medical Genetics in Britain: Laying the Foundation (1940–1960)." In *Encyclopedia of Life Sciences,* 1–5. Chichester, UK: John Wiley & Sons, 2009.

Index

Page numbers in italics refer to figures.

Printed in the USA
CPSIA information can be obtained
at www.ICGtesting.com
LVHW021047201023
761597LV00016B/1887